Bob Weil

Mapping Sciences Series

Editor-in-chief John G. Lyon

Titles include:

Aerial Mapping
Edgar Faulkner

Assessing the Accuracy of Remotely Sensed Data: Principles and Practices
Russell G. Congalton and Kass Green

DIGEST: A Primer for the International GIS Standard
Kelly Chan

Environmental GIS: Applications to Industrial Facilities
William J. Douglas

Practical Handbook for Wetland Identification and Delineation
John G. Lyon

Remote Sensing for Landscape Ecology: New Metric Indicators for Monitoring, Modeling, and Assessment of Ecosystems
Robert C. Frohn

Satellite Remote Sensing of Natural Resources
David L. Verbyla

Wetland and Environmental Applications of GIS
John G. Lyon and Jack McCarthy

Assessing *the* Accuracy *of* Remotely Sensed Data:
Principles and Practices

Russell G. Congalton
Kass Green

LEWIS PUBLISHERS
Boca Raton London New York Washington, D.C.

Library of Congress Cataloging-in-Publication Data

Congalton, Russell G., 1957–
Assessing the accuracy of remotely sensed data : principles and practices / Russell Congalton, Kass Green.
p. cm. — (Mapping science series)
Includes bibliographical references.
ISBN 0-87371-986-7 (alk. paper)
1. Remote sensing—Evaluation. I. Green, Kass. II. Title. III. Series.
G70.4.C647 1998
621.36′78—dc21   98-29658
                                                                                                                          CIP

© 1999 by CRC Press, Inc.
Lewis Publishers is an imprint of CRC Press

No claim to original U.S. Government works
International Standard Book Number 0-87371-986-7
Library of Congress Card Number 98-29658
Printed in the United States of America 1 2 3 4 5 6 7 8 9 0
Printed on acid-free paper

Dedication

This book is dedicated to Dr. Roy Mead for his vision, encouragement, advice, and commitment to assessing the accuracy of remotely sensed data.

About the Authors

Russell G. Congalton

Russell G. Congalton has spent much of the last 20 years developing techniques and practical applications for assessing the accuracy of remotely sensed data. This work began in 1979 as an MS student at Virginia Polytechnic Institute and State University, continued through his dissertation at the same institution, and has followed him throughout his academic career. Upon graduation, Dr. Congalton was employed as a post-doctorate research associate with the US Army Corps of Engineers Waterways Experiment Station Environmental Lab in 1984. From 1985-1991, he held the position of Assistant Professor of Remote Sensing in the Department of Forestry and Resource Management at the University of California, Berkeley. Also during this time he began his friendship with Kass Green and Pacific Meridian Resources which has lead to his current role as its Chief Scientist. Since 1991, Dr. Congalton has been on the faculty in the Department of Natural Resources at the University of New Hampshire. Currently he is an Associate Professor of Remote Sensing and GIS.

Dr. Congalton has published over 30 peer-reviewed articles and more than 40 conference proceedings papers. He is the author of three book chapters and is co-editor of a book on spatial uncertainty in natural resource databases. He has been a member of the American Society for Photogrammetry and Remote Sensing (ASPRS) since 1979. He was the Conference Director for GIS '87 in San Francisco and was the first National GIS Division Director serving on the National Board of Directors from 1989-91. Currently, he is the Principal Investigator for the Land Cover/Biology Investigation of the GLOBE Program, a project that integrates environmental science with K-12 education sponsored by NSF, NASA, and NOAA.

Ms. Kass Green

Kass Green is a cofounder and President of Pacific Meridian Resources, a natural resources, GIS and remote sensing consulting firm that operates from offices throughout the United States (www.pacificmeridian.com). Ms. Green's background includes over 25 years of experience in natural resource policy, economics, GIS analysis and remote sensing. She earned a BS in Forestry from the University of California, Berkeley, an MS in Natural Resource Policy and Economics from the University of Michigan, and a PhD–ABD from the University of California, Berkeley. Ms. Green has lead Pacific Meridian's growth since its inception in 1988 to its current status as a leader in remote sensing and GIS community. She is the author of numerous articles on GIS and remote sensing, regularly conducts workshops on the applications and practical uses of spatial data analysis, and is a highly requested speaker at many conferences and symposia.

Acknowledgments

No book is written solely by the authors listed on the cover and we have many people to thank who helped, encouraged, and inspired us along the way. First, we would like to thank the many graduate students at the University of New Hampshire that compiled, edited, and proofread this work along the way. Special thanks go to Dan Fehringer, Robb Macleod, and Lucie Plourde. Second, we would like to acknowledge the help of the entire staff at Pacific Meridian Resources especially Aaron Loeb, who spent countless hours formatting tables and figures. Third, we must thank all of our friends and colleagues in the remote sensing community who inspired and encouraged us on many occasions. We are especially thankful to Dr. John Jensen, Dr. Greg Biging, Dr. Tom Lillesand, Dr. Jim Smith, Mr. Ross Lunetta, Mr. Mike Renslow, and Dr. Jim Campbell for their positive feedback and support. This book would not have been written if not for Dr. John Lyon. He has been a constant source of incredible encouragement and has mentored us through this entire process. Finally, we would like to thank our families for the time they managed without us while we worked on this book.

Contents

Chapter 1: Introduction...1
Why Remote Sensing?..2
Why Accuracy Assessment?...3
Organization of the Book..4

Chapter 2: Overview..7
Aerial Photography...7
Digital Assessments..8

Chapter 3: Sample Design..11
How is the Map Information Distributed?....................................11
  The Classification Scheme...12
  The Distribution of Map Categories......................................14
    Discrete versus Continuous Data......................................14
    Spatial Autocorrelation..14
What is the Appropriate Sample Unit?..16
How Many Samples Should Be Taken?..17
  Binomial Distribution..19
  Multinomial Distribution..19
How Should the Samples Be Chosen?...22

Chapter 4: Data Collection...27
What Should Be the Source of the Reference Data?.......................28
  Using Existing versus Newly Collected Data..........................28
  Photos versus Ground..29
How Should the Reference Data Be Collected?.............................30
When to Collect Reference Data...35
Ensuring Objectivity and Consistency..36
  Data Independence..37
  Data Collection Consistency...38
  Quality Control..40

Chapter 5: Basic Analysis Techniques...43
Non-Site Specific Assessments..43
Site Specific Assessments...45
  The Error Matrix..45
    Mathematical Representation of the Error Matrix...............47
    Analysis Techniques..49
      Kappa..49
      Margfit..53
      Conditional Kappa...56
      Weighted Kappa..57
      Compensation for Chance Agreement............................58
      Confidence Limits...59
      Area Estimation/Correction...63

Chapter 6: Analysis of Differences in the Error Matrix..................**65**
Errors in the Reference Data...65
Sensitivity of the Classification Scheme to Observer Variability........................67
Inappropriateness of the Remote Sensing Technology...69
Mapping Error 69
Summary...70
Appendix 1 71

Chapter 7: Advanced Topics..**75**
Beyond the Error Matrix 75
  Modifying the Error Matrix...76
  Fuzzy Set Theory...76
  Measuring Variability..79
Complex Data Sets 80
  Change Detection...80
  Multilayer Assessments 83

Chapter 8: The California Hardwood Rangeland Monitoring Project........**85**
Introduction..85
Background..85
Sample Design...86
  How is the Map Information Distributed?...87
  What is the Appropriate Sample Unit?...89
  How Many Samples Should Be Taken?..91
  How Should the Samples Be Chosen and Distributed across the Landscape?...93
    Samples Chosen from the 1981 Coverage 93
    Samples Chosen from the 1990 Coverage 94
Data Collection 95
  What Should Be the Source Data for the Reference Samples? 95
  What Type of Information Should Be Collected? 96
  When Should the Reference Data Be Collected? 98
    Quality Control 98
Analysis..99
  Development of the Error Matrices..99
  Statistical Analysis..105
  Analysis of Off-Diagonal Samples..105
  Crown Closure Analysis 105
  Crown Closure Map Results...110
  Cover Type Analysis...114
  Cover Type Map Results 116
  Extent 120
Discussion...121
Conclusions..122

References...**125**

Index..**133**

CHAPTER 1

Introduction

As resources become scarce, they become more valuable. Value is evidenced both by the increasing prices of resources and by controversy over resource allocation and management. From forest harvesting and land use conversion throughout North America, to the fragmentation of tropical bird habitat, to acid rain deposition in Eastern Europe, to Siberian tiger habitat loss in Russia, people have significantly affected the ecosystems of the world. Expanding population pressures continue to cause the price of resources to increase and to intensify conflicts over resource allocation.

As resources become more valuable, the need for timely and accurate information about the type, quantity, and extent of resources multiplies. Allocating and managing the Earth's resources requires knowing the distribution of resources across space. To efficiently plan emergency response we need to know the location of roads relative to fire stations and police stations. To improve the habitat of endangered species such as the spotted owl or salmon, we need to know what the species habitat requirements are, where that habitat exists, where the animals exist, and how changes to the habitat and surrounding environments will affect species distribution and population viability. To plan for future developments we need to know where people will work, live, shop, and go to school. Because each decision (including the decision to do nothing) impacts (1) the status and location of resources and (2) the relative wealth of individuals and organizations who derive value from the resources, knowing the location of resources and how they interact spatially is critical to effectively managing those resources and ourselves.

Thus, decisions about resources require maps; and effective decisions require maps of known accuracy. For centuries, maps have provided important information concerning the distribution of resources across space. Maps help us to measure the extent and distribution of resources, analyze resource interactions, identify suitable locations for specific actions (e.g., development or preservation), and plan future events. If our decisions based upon map information are to have expected results, then the accuracy of the maps must be known. Otherwise, implementing decisions will result in surprises, and these surprises may be unacceptable. For example, if we have a map that displays forest, crops, urban, water, and barren land cover types, we can plan a picnic in the part of the forest that is near a lake. If we don't know the accuracy of the map, and the map is 100% accurate, we can travel to our forest location, and, in fact, find ourselves in a forest. However, if the maps are not 100% accurate we may find ourselves in the middle of the lake, when we were expecting a forest. However, if we know the accuracy of the map, we can incorporate the known expectations of accuracy into our planning and create contingency plans in situations when the accuracy is low. This type of knowledge is critical when we move from our lighthearted picnic example to more critical decisions such as endangered species preservation, resource allocation, peace-keeping actions, and emergency response.

The purpose of this book is to present the concepts of accuracy assessment and to teach readers how to adequately design and implement accuracy assessment procedures. The book concentrates on the classification accuracy of maps made from remotely sensed data.

WHY REMOTE SENSING?

Remote sensing is the collection and interpretation of information about an object from a remote vantage point. The most basic remote sensing devices are our own eyes and ears. Manmade systems include instruments on airplanes and satellites. Because there is a high correlation between variation in remotely sensed data and variation across the earth's surface, remotely sensed data provides an excellent basis for making maps of land use and land cover.

Remotely sensed data has captured our imaginations since the first camera was carried aloft in a hot air balloon. This fascination continues to grow with each new sensor developed and each new satellite launched. The "bird's eye view" offered by remotely sensed data is irresistible because it is a view that can be readily understood and is inimitably useful, yet it is impossible to obtain without the help of technology.

From the advent of the first aerial photographs to the launch of the latest satellite imaging system, remotely sensed data has become an increasingly important and efficient way of collecting map information. We use remotely sensed data to make maps because it

- Is usually less expensive and faster than creating maps from information collected on the ground,
- Offers a perspective from above (the "bird's eye view"), allowing for a better understanding of spatial relationships, and
- Permits capturing types of data that humans can't sense, such as the infrared portions of the electromagnetic spectrum.

The accuracy of maps made from remotely sensed data is measured by two types of criteria: location accuracy and classification or thematic accuracy. Location accuracy refers to how precisely map items are located on the map relative to their true location on the ground. Thematic accuracy refers to the accuracy of the map label in describing a class or condition on the earth. For example, in the picnic example, the earth's surface was classified as either forest, water, crops, urban, or barren. In

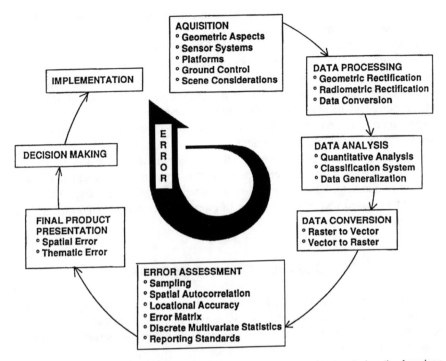

Figure 1-1 Sources of error in remotely sensed data. Reproduced with permission, the American Society for Photogrammetry and Remote Sensing, from: Lunetta, R., R. Congalton, L. Fenstermaker, J. Jensen, K. McGwire, and L. Tinney. 1981. Remote sensing and geographic information system data integration: error sources and research issues. *Photogrammetric Engineering and Remote Sensing.* Vol. 57, No. 6, pp. 677-687.

thematic map accuracy we are interested in whether or not the lake has been accurately labeled water or inaccurately labeled forest. We want to estimate the probable types and magnitude of label confusion across the entire map.

The widespread acceptance and use of remotely sensed data has been and will continue to be dependent on the quality of the map information derived from it. However, map inaccuracies or error can occur at many steps throughout any remote sensing project. Figure 1-1 shows a schematic diagram of the many possible sources of error. Accuracy assessment is conducted to understand the quality of map information by identifying and assessing map errors.

WHY ACCURACY ASSESSMENT?

There are many reasons for performing an accuracy assessment. Perhaps the simplest reason is curiosity—the desire to know how good something is. In addition to the satisfaction gained from this knowledge, we also need to increase the quality of the map information by identifying and correcting the sources of errors. Third, analysts often need to compare various techniques, algorithms, analysts, or interpreters to test which is best. Finally, if the information derived from the remotely sensed data is to be used in some decision-making process, then it is critical that some measure of its quality be known.

Accuracy assessment determines the quality of the information derived from remotely sensed data. Accuracy assessment can be qualitative or quantitative, expensive or inexpensive, quick or time-consuming, well-designed and efficient or haphazard. The purpose of *quantitative* accuracy assessment is the identification and measurement of map errors. Quantitative accuracy assessment involves the comparison of a site on a map against reference information for the same site. The reference data is assumed to be correct.

Usually funding limitations preclude the assessment of every spatial unit on the map. Because comparison of every spatial point is impractical, sample comparisons are used to estimate the accuracy of maps. Accuracy assessment requires (1) the design of unbiased and consistent sampling procedures, and (2) rigorous analysis of the sample data.

ORGANIZATION OF THE BOOK

More and more researchers, scientists, and users are discovering the need to adequately assess the results of remotely sensed data. Still, at this time there are many more questions about accuracy assessment than there are answers. This book addresses a few of the most important ones:

1. How many accuracy assessment samples should be collected and how should these samples be allocated across the map?
2. What sampling schemes should be used to select accuracy assessment samples?
3. What types of reference data should be collected? Are aerial photographs appropriate, or must ground observations or measurements be made?
4. How should the accuracy assessment samples be analyzed?
5. What is an error matrix and how should it be used?
6. What are the statistical properties associated with the error matrix and what analysis techniques are applicable?
7. What other techniques beyond the error matrix exist to aid in accuracy assessment?

The objective of this book is to provide the reader with the principles and practical considerations of designing and conducting accuracy assessment. All accuracy assessments include four fundamental steps:

- Designing the sample
- Collecting data for each sample
- Building and testing the error matrix
- Analyzing the results

Each step must be rigorously planned and implemented. First, sample areas of the map are selected, and information is collected both from the map and from the reference data for each sample. Next, the map and reference labels are compared to one another in an error matrix. Finally, the results of the matrix are analyzed for statistical significance and for reasonableness. Effective accuracy assessment requires (1) design and implementation of unbiased sampling procedures, (2) consistent and accurate collection of sample data, and (3) rigorous comparative analysis of the sample data.

The organization of this book takes you through each of these fundamental steps. Chapter 2 begins with a discussion of the history and basic concepts of accuracy assessment. Chapter 3 introduces sampling design. Chapter 4 is devoted to the collection of reference site data. Chapter 5 reviews the fundamental concepts of accuracy assessment analysis including a discussion of site versus non-site specific methods, error matrix generation, confidence intervals, area estimation, and statistical analysis techniques. Chapter 6 reviews the steps for analyzing accuracy assessment results. Chapter 7 is devoted to discussing advanced topics, including fuzzy logic and the assessment of multilayered and multitemporal map layers. The final chapter summarizes the book with the presentation of a real-world case study.

CHAPTER 2

Overview

The history of remote sensing is a relatively short one. Aerial photography has been used as a mapping tool effectively for just over half a century. Digital image scanners and cameras on satellites and airplanes have even a shorter history spanning a little over 2 decades. However, it was the development of these digital devices that had the most profound impact on accuracy assessment for all remotely sensed data.

AERIAL PHOTOGRAPHY

The first aerial photograph was acquired from a balloon in 1858, but it wasn't until 1909 that Wilbur Wright took the first aerial photograph from an airplane. Even then the extensive use of aerial photography for mapping and interpreting land use and land cover didn't begin until after World War II (Spurr 1948). In these early days, the focus was primarily on the development of new cameras and other instruments to make the best use of the aerial photographs. Spurr, in his excellent book entitled *Aerial Photographs in Forestry* (1948), presents the prevailing opinion about assessing the accuracy of photo interpretation. He states, "Once the map has been prepared from the photographs, it must be checked on the ground. If preliminary reconnaissance has been carried out, and a map prepared carefully from good quality photographs, ground checking may be confined to those stands whose classification could not be agreed upon in the office, and to those stands passed through en route to these doubtful stands." In other words, a qualitative visual check to see if the map looks right has traditionally been recommended.

Finally, in the 1950s some researchers saw the need for quantitative assessment of photo interpretation in order to promote their discipline as a science (Sammi 1950, Katz 1952, Young 1955, Colwell 1955). In a panel discussion entitled, "Reliability of Measured Values" held at the 18th Annual Meeting of the American Society of Photogrammetry, Mr. Amrom Katz (1952), the panel chair, made a very compelling plea for the use of statistics in photogrammetry. Other panel discussions were held and talks were presented that culminated with a paper by Young and Stoeckler (1956). In this paper they actually propose techniques for a quantitative evaluation of photo interpretation, including the use of an error matrix to compare field and photo classifications, and a discussion of the boundary error problem.

Unfortunately, these techniques never received widespread attention nor acceptance. The *Manual of Photo Interpretation* published by the American Society of Photogrammetry (1960) does mention the need to train and test photo interpreters. However, it contains no description of the quantitative techniques proposed by a brave few in the 1950s. Nor will the reader find a textbook today on photo interpretation that deals with these techniques.

It seems that photo interpretation has become a time-honored skill, and the prevailing opinion is that a quantitative assessment is unnecessary. In speaking with some of the old-time photo interpreters, they remember those times when quantitative assessment was an issue. In fact, they mostly agree with the need to perform such an assessment and are usually the first to point out the limitations of photo interpretation.

And so the quantitative assessment of photo interpretation has never become a requirement of any project. Rather, the assumption that the map was correct or at least good enough prevailed. Then along came digital remote sensing, and some of these fundamental assumptions about photo interpretation were shaken.

DIGITAL ASSESSMENTS

As in the early days of aerial photography, the launch of Landsat 1 in 1972 resulted in a great burst of exuberant effort as researchers and scientists charged ahead trying to develop the field of digital remote sensing. In those early days, much progress was made and there wasn't much time to sit back and evaluate how they were doing. After approximately 5 years into the Landsat program, researchers began to consider and realistically evaluate where they were going and to some extent how they were doing with respect to quality.

The history of accuracy assessment of remotely sensed data is relatively short, beginning around 1975. Researchers, notably Hord and Brooner (1976), van Genderen and Lock (1977), and Ginevan (1979), proposed criteria and techniques for testing map accuracy. In the early 1980s more in-depth studies were conducted and new techniques proposed (Rosenfield et al., 1982; Congalton et al., 1983; and Aronoff, 1985). Finally, from the late 1980s up to the present time, a great deal of work has been conducted on accuracy assessment. More and more researchers, scientists, and users are discovering the need to adequately assess the results of remotely sensed data.

The history of digital accuracy assessment can be effectively divided into four parts or epochs. Initially, no real accuracy assessment was performed, but rather an "it looks good" mentality prevailed. This approach is typical of a new, emerging technology in which everything is changing so quickly that there is not time to sit back and assess how well you are doing. Despite the maturing of the technology over the last 15 years, some remote sensing analysts are still stuck in this mentality.

The second epoch is called the age of non-site specific assessment. During this period, overall acreages were compared between ground estimates and the map

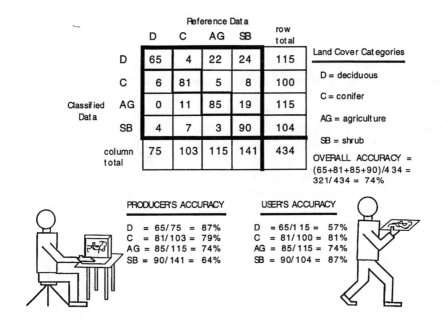

Figure 2-1 Example error matrix.

without regard for location. The second epoch was relatively short-lived and quickly led to the age of site specific assessments.

In a site specific assessment, actual places on the ground are compared to the same place on the map and a measure of overall accuracy (i.e., percent correct) presented. Site specific assessment techniques were the dominant method until the late 1980s.

Finally, the fourth and current age of accuracy assessment could be called the age of the error matrix. An error matrix compares information from reference sites to information on the map for a number of sample areas. The matrix is a square array of numbers set out in rows and columns that express the labels of samples assigned to a particular category in one classification relative to the labels of samples assigned to a particular category in another classification (Figure 2-1). One of the classifications, usually the columns, is assumed to be correct and is termed the *reference data*. The rows usually are used to display the *map labels* or classified data generated from the remotely sensed data. Thus, two labels from each sample are compared to one another:

- Reference data labels: the class label of the accuracy assessment site derived from data collected that is assumed to be correct; and
- Classified data or map labels: the class label of the accuracy assessment site derived from the map.

Error matrices are very effective representations of map accuracy, because the individual accuracies of each map category are plainly described along with both the errors of inclusion (commission errors) and errors of exclusion (omission errors) present in the map. A commission error occurs when an area is included in an incorrect category. An omission error occurs when an area is excluded from the category to which it belongs.

In addition to clearly showing errors of omission and commission, the error matrix can be used to compute overall accuracy, producer's accuracy, and user's accuracy (Story and Congalton 1986). Overall accuracy is simply the sum of the major diagonal (i.e., the correctly classified pixels or samples) divided by the total number of pixels or samples in the error matrix. This value is the most commonly reported accuracy assessment statistic. Producer's and user's accuracies are ways of representing individual category accuracies instead of just the overall classification accuracy (see Chapter 5 for more details on the error matrix).

Proper use of the error matrix includes correctly sampling the map and rigorously analyzing the matrix results. The techniques and considerations of the building and analyzing of an error matrix are the main themes of this book.

CHAPTER 3

Sample Design

Accuracy assessment requires sampling, and sampling requires design of the distribution and types of samples to be taken, and collection of data from the sample areas. The selection of a proper and efficient sample design to collect valid reference data is one of the most challenging and important components of any accuracy assessment because the design will determine both the cost and the statistical rigor of the assessment.

Because accuracy assessment assumes that the information displayed in the error matrix is a true representation of the map being assessed, an improperly gathered sample will produce meaningless results. Several considerations are critical to designing an accuracy assessment sample that is truly representative of the map:

1. How is the map information distributed?
2. What is the appropriate sample unit?
3. How many samples should be taken?
4. How should the samples be chosen?

While seemingly straightforward, each of these steps has potential pitfalls. Failure to consider even one of them can lead to serious shortcomings in the assessment process. This chapter considers each one of these issues.

HOW IS THE MAP INFORMATION DISTRIBUTED?

How we sample the map for accuracy will partially be driven by how the information on the map is distributed across space by map category. This distribution will, in turn, be a function of how we have chosen to categorize the features of the earth being mapped—referred to as the *classification scheme*. Once we know the classification scheme, we can then learn more about how map classes are distributed. Important considerations are the discrete nature of map information and the spatial interrelationship or autocorrelation of that information. Assumptions made about the distribution of map categories will affect both how we select accuracy assessment samples and the outcome of the analysis.

The Classification Scheme

Maps categorize the earth's surface. For example, road maps tell us the type of road, its name, and location. Land-cover maps typically enumerate the types of vegetation covering the earth, such as trees, shrubs, grass, or barren.

Map categories are specified by the project's classification scheme. Classification schemes are a means of organizing spatial information in an orderly and logical way (e.g., Cowardin et al. 1979). Classification schemes are fundamental to any mapping project because they create order out of chaos and reduce the total number of objects (i.e., classes) that we must deal with to some reasonably small number. The classification scheme makes it possible for the map producer to characterize landscape features and for the user to readily recognize them. The detail of the scheme is driven by (1) the anticipated uses of the map information, and (2) the features of the earth that can be discerned with the data (e.g., aerial photography, satellite imagery) being used to create the map. If a rigorous classification scheme is not developed before mapping begins, then any subsequent accuracy assessment of the map will be meaningless because it will be impossible to definitively state that an accuracy assessment sample area is of one class or another.

A classification scheme has two critical components: (1) a set of *labels* (urban residential, deciduous forest, etc.), and (2) a set of *rules* or definitions such as a dichotomous key—or a system—for assigning labels (e.g., a "deciduous forest must have at least 75% crown closure in deciduous trees"). Without a clear set of rules, the assignment of labels to types can be arbitrary and lack consistency. For example, everyone has their own idea about what constitutes a forest, yet there are many definitions that could result in very different maps of forest distribution. The U.S. Forest Service uses the definition that an area is called forest if 10% of the ground area is covered by trees. The U.S. Environmental Protection Agency (EPA) has used a broader definition in which forest exists only if 25% of the ground area is covered by trees.

The level of detail (i.e., number and complexity of the categories) in the scheme strongly influences the time and effort needed to conduct the accuracy assessment. The more detailed the scheme, the more expensive the map and its assessment. Because the classification scheme is so important, no work should begin on a mapping project until the scheme has been thoroughly reviewed and as many problems as possible identified and solved.

In addition to having labels and a set of rules, a classification scheme should be (1) *mutually exclusive* and (2) *totally exhaustive*. Mutual exclusivity requires that each mapped area fall into one and only one category or class. For example, the rules of the classification scheme will need to clearly distinguish between forest and water (seemingly simple) so that a mangrove swamp cannot receive both a forest and a water label. A totally exhaustive classification scheme results in every area on the mapped landscape receiving a map label; no area can be left unlabeled.

Table 3-1   Example Classification Scheme

Rules   *Labels*

If water is ≥ 70% of the site, then ....Water

Or

If fuel vegetation is ≤ 29% of the site, then ....Non-fuel

Or

If fuel vegetation is ≥ 30% of the site, and

   If fuel vegetation cover is ≥ 50% shrub, then...............Heavy fuel

   Or

   If fuel vegetation cover is ≤ 49% shrub and

   ≥ 50% trees, then...Medium fuel

If fuel vegetation cover is ≤ 49% shrub or trees, then..........Light fuel

Note: Fuel vegetation is defined as growing or dead vegetation capable of carrying a wildfire. Crops, golf courses, irrigated parks, and wetlands are explicitly excluded.

Finally, if possible, it's also advantageous to use a classification scheme that is *hierarchical*. In hierarchical systems, specific categories within the classification scheme can be collapsed to form more general categories. This ability is especially important when it is discovered that certain map categories cannot be reliably mapped. For example, it may be impossible to separate interior live oak from canyon live oak in California's oak woodlands (the types are almost indistinguishable on the ground). Therefore, these two categories may have to be collapsed to form a live oak category that can be reliably mapped. Table 3-1 provides a simple example classification scheme for a fuel mapping project.

It is critical that reference data be collected and labeled using the same classification scheme as was used to generate the map. This may seem obvious until you are tempted to use an existing map to assess the accuracy of a new map. Rarely will two maps have been made using the same classification scheme. For example, if numerous federal agencies are working together it is possible for one agency to map forest as any area on the ground with 10% or more of tree cover. Another agency may be collecting the reference data, and in their classification scheme a forest is defined as any area on the ground with 25% or more of tree cover. Clearly, using these reference data to assess the map will result in concluding that the map is not very good (i.e., that the map has included a lot of forest areas that are not forest according to the reference data) when in fact the two maps differed because the classification schemes were different.

The Distribution of Map Categories

Most statistics assume that the population to be sampled is continuous and normally distributed, and that samples will be independent. Yet we know that classification systems, for all their power in organizing chaos, also take a continuous landscape and divide it into often arbitrary discrete categories. For example, tree crown closure rarely falls in discrete classes. Yet when we make a map of crown closure, we impose discrete crown closure classes across the landscape. If we have a boundary between two crown closure classes at 70% crown closure, we can expect to find confusion between a forest stand with a crown closure of 67% that belongs in Class 3 and a stand of 73% that belongs in Class 4. In addition, categories tend to be related spatially, resulting in autocorrelation. In most situations, some balance between what is statistically valid and what is practically obtainable is desired. Therefore, knowledge of these statistical considerations is a must.

Discrete versus Continuous Data

Most students who have completed a beginning statistics course are familiar with sampling and analysis techniques for continuous, normally distributed data. It is these techniques, such as analysis of variance (ANOVA) and linear regression, that are most familiar to the reader.

However, map information is discrete, not continuous, and frequently not normally distributed. Therefore, normal theory statistical techniques that assume a continuous distribution may be inappropriate for map accuracy assessment. It is important to consider how the data are distributed and what assumptions are being made before performing any statistical analysis. Sometimes there is little that can be done about the artificial delineations in the classification scheme; other times the scheme can be modified to better represent natural breaks.

Spatial Autocorrelation

Spatial autocorrelation is said to occur when the presence, absence, or degree of a certain characteristic affects the presence, absence, or degree of the same characteristic in neighboring units (Cliff and Ord 1973), thereby violating the assumption of sample independence. This condition is particularly important in accuracy assessment if an error in a certain location can be found to positively or negatively influence errors in surrounding locations (Campbell 1981).

The existence of spatial autocorrelation is clearly illustrated in work by Congalton (1988a) on Landsat MSS data from three areas of varying spatial diversity/complexity (i.e., an agriculture, a rangeland, and a forested site), which showed a positive influence over 1 mile away. Figure 3-1 presents the results of this analysis. Each

AGRICULTURE   RANGELAND   FOREST

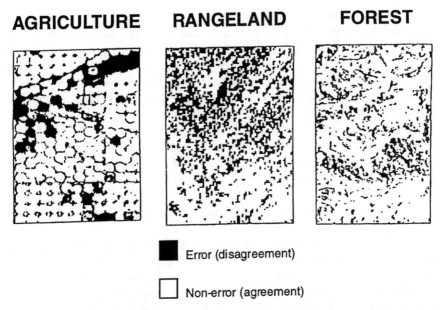

■ Error (disagreement)

☐ Non-error (agreement)

Figure 3-1 Difference images (7.5 minute quads) showing the pattern of error for three eco-systems of varying complexity: agriculture, rangeland, and forest. Reproduced with permission, the American Society for Photogrammetry and Remote Sensing, from: Congalton, R. 1988. Using spatial autocorrelation analysis to explore errors in maps generated from remotely sensed data. *Photogrammetric Engineering and Remote Sensing*. Vol. 54, No. 5, pp. 587-592.

image, called a difference image, is a comparison between the remotely sensed classification and the reference data. The black areas represent the error, those places where the remotely sensed classification and the reference data disagree, and the white areas represent the agreement.

These results are readily explainable in an agricultural environment where field sizes are large and typical misclassification would result in an error in labeling the entire field. In the agricultural difference image in Figure 3-1, the fields are center pivot irrigation, circular fields, and examples can be seen of misclassifying entire fields.

However, these results are more surprising for the rangeland and forested sites. Both sites are more spatially complex than the agriculture site. The rangeland area has some of the wide open fields similar to agriculture and some of the edge effects more common to the complex forest site.

The forest site is the most spatially complex, and the error would be expected to occur mostly in the edges or transition zones between forest types. While viewing the forest difference image does tend to confirm these edge problems, the results of the analysis still indicate that there is strong positive autocorrelation up to 30 pixels away. In other words, an error at a given location means that it is more likely to find another error within this rather large distance away (i.e., 30 MSS pixels) than it is to find agreement.

Surely these results should affect the sample size and especially the sampling scheme used in accuracy assessment, especially in the way this autocorrelation affects the assumption of sample independence. This autocorrelation may then be responsible for periodicity in the data that could affect the results of any type of systematic sample. In addition, the size of the cluster used in cluster sampling would also be affected because each sample would not be contributing new, independent information, but rather redundant information. Therefore, it would not be practical to collect information in a large cluster sample, since very quickly each new polygon sampled would be adding very little new information.

However, cluster samples are cost-effective, especially in the field, when the cost of traveling from one sample to another can be very high. Even when the accuracy assessment samples are taken from aerial photography, clustering samples can create savings in set-up time for each photo.

WHAT IS THE APPROPRIATE SAMPLE UNIT?

Sample units are the portions of the map that will be sampled for accuracy assessment. There are four choices: a single pixel, a cluster of pixels (often a 3 × 3 pixel square), a polygon, and a cluster of polygons. Regardless of the sample unit, reference data should be collected at the same minimum mapping unit as was applied to the map generated from the remotely sensed data. Failure to consider this issue can cause huge problems. It is not possible to assess the accuracy of a 30 m × 30 m pixel with one 1/20 hectare plot. Nor is it possible to assess the accuracy of an AVHRR 1.1 km × 1.1 km pixel using a 30 m × 30 m pixel. It is critical that a true representation of both the map and the reference data be selected for a valid assessment.

Historically, a single pixel has been a poor choice for many reasons. First, it is inappropriate for maps created from aerial photography, because photos are not collected in pixels. Second, a pixel is an arbitrary rectangular delineation of the landscape that may have little relation to the actual delineation of land cover type. Third, before the relatively new geocoding and terrain correction procedures, it was almost impossible to exactly align one pixel on an image to the exact same area in the reference data. Therefore, there was no way to guarantee that you were looking at the same pixel on the image as you were on the reference data. Finally, until global positioning systems (GPS) came along there was no good way to precisely locate yourself on the ground to assure you were collecting reference data for the correct pixel.

A cluster of pixels, typically a 3 × 3 box, has been the most common choice for the sample unit. A cluster minimizes registration problems because it is larger than 1 pixel and therefore easier to locate on the reference data. However, a cluster of pixels (especially a 3 × 3 window) may still be an arbitrary delineation of the landscape, resulting in the sample unit encompassing more than one map category. To avoid this problem, some researchers require that only homogeneous clusters of pixels be sampled. However, such restrictions may result in a biased sample that avoids heterogeneous areas that are a function of a mix of pixels (e.g., a mixed hardwood conifer stand of trees), as depicted in Figure 3-2. It is important to remember that the sample unit dictates the level of detail for the accuracy assessment. If the assessment is performed on a 3 × 3 cluster of pixels, then nothing can be said about an individual pixel; nor can anything be said about polygons (management areas, forest stands, agricultural fields, etc.).

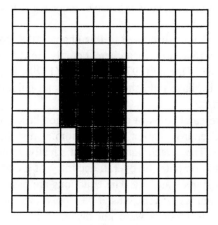

Homogeneous polygon

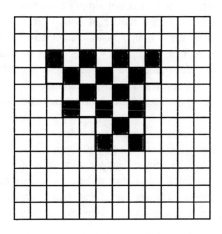

Heterogeneous polygon

Figure 3-2 Comparison of accuracy assessment polygons comprised of homogeneous versus heterogeneous pixels.

If the objective of the mapping is to produce a polygon map for use in wildlife or ecosystem management, it is important that the assessment be conducted on the polygon basis. The resulting accuracy values then tell the user and producer of the map about the level of detail they are interested in–the polygons. More and more mapping projects from remotely sensed data are generating polygon products, and therefore the polygon is replacing the cluster of pixels as the sample unit of choice.

HOW MANY SAMPLES SHOULD BE TAKEN?

Accuracy assessment requires that an adequate number of samples per map class be gathered so that any analysis performed is statistically valid. However, the collection of data at each sample point is very expensive, requiring that sample size be kept to a minimum to be affordable.

Of all the considerations discussed in this chapter, the most has probably been written about sample size. Many researchers, notably Hord and Brooner (1976), van Genderen and Lock (1977), Hay (1979), Ginevan (1979), Rosenfield et al. (1982), and Congalton (1988b), have published equations and guidelines for choosing the appropriate sample size.

The majority of researchers have used an equation based on the binomial distribution or the normal approximation to the binomial distribution to compute the required sample size. These techniques are statistically sound for computing the sample size needed to compute the overall accuracy of a classification or even the overall accuracy of a single category. The equations are based on the proportion of correctly classified samples (pixels, clusters, or polygons) and on some allowable error. However, these techniques were not designed to choose a sample size for filling in an error matrix.

In the case of an error matrix, it is not simply a question of correct or incorrect (the binomial case). Instead, it is a matter of which error or which categories are being confused. Given an error matrix with n land cover categories, for a given category there is one correct answer and $(n - 1)$ incorrect answers. Sufficient samples must be acquired to be able to adequately represent this confusion. Therefore, the use of the binomial distribution for determining the sample size for an error matrix is not appropriate. Instead, the use of the multinomial distribution is recommended (Tortora 1978).

Because of the large number of pixels in a remotely sensed image, traditional thinking about sampling often does not apply. Even as small a sample as one half percent of a single Landsat Thematic Mapper scene can be over 300,000 pixels. As already discussed, most accuracy assessments are not performed on a per pixel basis, but the same relative argument holds true if the sample unit is a cluster of pixels or a polygon. Therefore, practical considerations more often dictate the sample size selection.

A balance between what is statistically sound and what is practically attainable must be found. In our experience, a general guideline or good "rule of thumb" seems to be collecting a minimum of 50 samples for each vegetation or land cover category in the error matrix. This guideline was empirically derived over many projects. However, the use of the multinomial equation has confirmed that it is a good balance between statistical validity and practicality.

Of course, the appropriate sample size can and should be computed for each project using the multinomial distribution. However, this "rule of thumb" can be used as a general starting point and for initial planning purposes. The general guidelines further state that if the area is especially large (i.e., more than a million acres) or the classification has a large number of vegetation or land cover categories (i.e., more than 12 categories), the minimum number of samples should be increased to 75 or 100 samples per category.

The number of samples for each category can also be adjusted based on the relative importance of that category within the objectives of the mapping or by the inherent variability within each of the categories. Sometimes it is better to concentrate the sampling on the categories of interest and increase their number of samples while reducing the number of samples taken in the less important categories. Also it may be useful to take fewer samples in categories that show little variability such as water or forest plantations and increase the sampling in the categories that are more variable such as uneven-aged forests or riparian areas. However, in most instances, some minimum number of samples should be taken in each land cover category contained in the matrix.

Obviously, the sample size could be designed to select many samples in categories that are most accurate and few in the confused categories. This result may not be representative of the map accuracy, but would guarantee a high accuracy value. Care should be taken to assure that the sampling effort is carefully planned and implemented. It should also be noted that exactly how the sample is selected can affect the analysis performed on the sampled data. Again, the object here is to balance the statistical recommendations in order to get an adequate sample to generate an appropriate error matrix with the time, cost, and practical limitations associated with any viable remote sensing project.

Binomial Distribution

As mentioned above, the binomial distribution or the normal approximation to the binomial distribution is appropriate for computing the sample size for determining overall accuracy or the accuracy of an individual category. It is appropriate for the two-case situation where only right and wrong are important. Choosing the appropriate sample size from the binomial or normal approximation is dependent on (1) the level of acceptable error one is willing to tolerate and (2) the desired level of confidence that the actual accuracy is within some minimum range. Numerous publications present look-up tables of the required sample size for a given acceptable error and desired level of confidence (e.g., Cochran 1977 and Ginevan 1979).

For example, suppose it is decided that a map is unacceptable if the overall accuracy is 90% or less. Also, let's say that we are willing to accept a 1 in 20 chance that we will make a mistake based on our sample and accept a map that actually has an accuracy of less than 90%. Finally, let us decide that we will accept the same risk, a 1 in 20 chance, of rejecting a map that is actually correct. The appropriate look-up table would then indicate that we must take 298 samples of which only 21 can be misclassified. If more than 21 samples were misclassified, we would conclude that the map is not acceptable.

Multinomial Distribution

As discussed earlier in this chapter, the multinomial distribution provides the appropriate equations for sampling to build an error matrix. The procedure for generating the appropriate sample size from the multinomial distribution is summarized here and was originally presented by Tortora (1978).

Consider a population of units divided into k mutually exclusive and exhaustive categories. Let Π_i, $i = 1, ..., k$, be the proportion of the population in the ith category, and let n_i, $i = 1, ..., k$, be the frequency observed in the ith category in a simple random sample of size n from the population.

For a specified value of α, we wish to obtain a set of intervals S_i, $i = 1, ..., k$, such that

$$\Pr\left\{\bigcap_{i=1}^{k}(\Pi_i \in S_i)\right\} \geq 1 - \alpha.$$

That is, we require the probability that every interval S_i contains Π_i to be at least $1 - \alpha$. Goodman (1965) determined the approximate large-sample confidence interval bounds (when $n \to \infty$) as

$$\Pi_i^- \leq \Pi_i \leq \Pi_i^+,$$

where

$$\Pi_i^- = \Pi_i - \left[B\Pi_i(1 - \Pi_i)/n\right]^{1/2}, \tag{3.1}$$

$$\Pi_i^+ = \Pi_i + \left[B\Pi_i(1 - \Pi_i)/n\right]^{1/2}, \tag{3.2}$$

and B is the upper $(\alpha/k) \times 100$th percentile of the χ^2 distribution with 1 degree of freedom. These equations are based on Goodman's (1965) procedure for simultaneous confidence interval estimation.

Examining Equations (3.1) and (3.2), we see that $\left[\Pi_i(1 - \Pi_i)/n\right]^{1/2}$ is the standard deviation for the ith cell of the multinomial population. Also, it is important to realize that each marginal probability mass function is binomially distributed. If N is the total population size, then using the finite population correction factor (fpc) and the variance for each Π_i (from Cochran 1977), the approximate confidence bounds are

$$\Pi_i^- = \Pi_i - \left[B(N - n)\Pi_i(1 - \Pi_i)/(N - 1)n\right]^{1/2}, \tag{3.3}$$

$$\Pi_i^+ = \Pi_i + \left[B(N - n)\Pi_i(1 - \Pi_i)/(N - 1)n\right]^{1/2}. \tag{3.4}$$

Note that as $N \to \infty$ (3.3) and (3.4) converge to (3.1) and (3.2), respectively.

Next, in order to determine the required sample size, the precision for each parameter in the multinomial population must be specified. If the absolute precision for each cell is set to b_i, then (3.1) and (3.2) become

$$\Pi_i - b_i = \Pi_i - \left[B\Pi_i(1 - \Pi_i)/n\right]^{1/2}, \tag{3.5}$$

$$\Pi_i + b_i = \Pi_i + \left[B\Pi_i(1 - \Pi_i)/n \right]^{1/2}, \tag{3.6}$$

respectively. Similar results are obtained when the fpc is included. Equations (3.5) and (3.6) can be rearranged to solve for b_i (the absolute precision of the sample):

$$b_i = \left[B\Pi_i(1 - \Pi_i)/n \right]^{1/2}. \tag{3.7}$$

Then by squaring this Equation (3.7) and solving for n, the result is

$$n = B\Pi_i(1 - \Pi_i)/b_i^2, \tag{3.8}$$

or, using the fpc,

$$n = BN\Pi_i(1 - \Pi_i)/\left[b_i^2(N-1) + B\Pi_i(1 - \Pi_i) \right]. \tag{3.9}$$

Therefore, one should make k calculations to determine the sample size, one for each pair (b_i, Π_i), $i = 1, \ldots, k$, and select the largest n as the desired sample size. As functions of Π_i and b_i, Equations (3.8) and (3.9) show that n increases as $\Pi_i \to \frac{1}{2}$ or $b_i \to 0$.

In rare cases, a relative precision b_i' could be specified for each cell in the error matrix and not just each category. Here $b_i = b_i'\Pi_i$. Substituting this into Equation (3.8) gives

$$n = B(1 - \Pi_i)/\Pi_i b_i'^2. \tag{3.10}$$

A similar sample size calculation including the fpc can be computed as above.

Here again, one should make k calculations, one for each pair (b_i', Π_i), $i = 1, \ldots, k$. The largest n computed is selected as the desired sample size. As $\Pi_i \to \frac{1}{2}$ or $b_i' \to 0$, the sample size increases according to (3.10). If $b_i' = b'$ for all i, then the largest sample size is $n = B(1 - \Pi)/\Pi b'^2$, where $\Pi = \min(\Pi_1, \ldots, \Pi_k)$.

In the majority of cases for assessing the accuracy of remotely sensed data, an absolute precision is set for the entire classification and not each category or each cell. Therefore, $b_i = b$ and the only sample size calculation required is for the Π_i closest to $\frac{1}{2}$. If there is no prior knowledge about the values of the Π_i's, a "worst" case calculation of sample size can be made assuming some $\Pi_i = \frac{1}{2}$ and $b_i = b$ for $i = 1, \ldots, k$. In this worst-case scenario, the sample size required to generate a valid error matrix can be obtained from this simple equation as follows:

$$n = B/4b^2.$$

This approach can be made much clearer with a numerical example. First let's look at an example using the full Equation (3.8) and then at the corresponding sample size using the worst case or conservative sample size equation. Assume that there are eight categories in our classification scheme ($k = 8$), that the desired confidence level is 95%, that the desired precision is 5%, and that this particular class makes up 30% of the map area ($\Pi_i = 30\%$). The value for B must be determined from a Chi square table with 1 degree of freedom and $1 - \alpha/k$. In this case the appropriate value for B is $\chi^2_{(1,0.99375)} = 7.568$. Therefore, the calculation of the sample size is as follows:

$$n = B\Pi_i(1 - \Pi_i)/b_i^2$$

$$n = 7.568(0.30)(1 - 0.30)/(0.05)^2$$

$$n = 1.58928/0.0025$$

$$n = 636.$$

A total of 636 samples should be taken to adequately fill an error matrix or approximately 80 samples per class given that there were 8 classes in this map.

If the simplified, worst-case scenario equation is used, then the class proportion is assumed to be 50% and the calculation is as follows:

$$n = B/4b^2$$

$$n = 7.568/4(0.05)^2$$

$$n = 7.568/0.01 = 757.$$

In this worst case scenario, approximately 95 samples per class or 757 total samples would be required.

If the confidence interval is relaxed from 95% to 85%, the required sample sizes decrease. In the example above, the new appropriate value for B would be $\chi^2_{(1,0.98125)}$ = 5.695 and the total samples required would be 478 and 570 for the complete equation and the worst-case scenario, respectively.

HOW SHOULD THE SAMPLES BE CHOSEN?

In addition to the considerations already discussed, the choice and distribution of samples, or sampling scheme, is an important part of any accuracy assessment. Selection of the proper scheme is critical to generating an error matrix that is representative of the entire map. First, to reach valid conclusions about a map's accuracy from some sample of that map, the sample must be selected without bias. Failure to meet this important criteria affects the validity of any further analysis performed using the data because the resulting error matrix may over- or underestimate the true accuracy. Second, further data analysis will depend on which sampling scheme is selected. Different sampling schemes assume different sampling models and consequently, different variance equations to compute the required accuracy methods. Finally, the sampling scheme will determine the distribution of samples across the landscape, which will significantly affect accuracy assessment costs.

Many researchers have expressed opinions about the proper sampling scheme to use (e.g., Hord and Brooner 1976, Rhode 1978, Ginevan 1979, Fitzpatrick-Lins 1981, and Stehman 1992). These opinions vary greatly among researchers and include everything from simple random sampling to stratified, systematic, unaligned sampling.

There are five common sampling schemes that have been applied for collecting reference data: simple random sampling, systematic sampling, stratified random sampling, cluster sampling, and stratified systematic unaligned sampling. In a simple random sample each sample unit in the study area has an equal chance of being selected. In most cases, a random number generator is used to pick random x,y coordinates to go and collect samples. The main advantage of simple random sampling is the good statistical properties that result from the random selection of samples.

Systematic sampling is a method in which the sample units are selected at some equal interval over the study area. In most cases, the first sample is randomly selected and each successive sample is taken at some specified interval thereafter. The major advantage of systematic sampling is the ease in sampling somewhat uniformly over the entire study area.

Stratified random sampling is similar to simple random sampling; however, some prior knowledge about the study area is used to divide the area into groups or strata, and then each strata is randomly sampled. In the case of accuracy assessment, the map has been stratified into land cover or vegetation types. The major advantage of stratified random sampling is that all strata (i.e., land cover types), no matter how small, will be included in the sample.

In addition to the sampling schemes already discussed, cluster sampling has also been frequently used in assessing the accuracy of remotely sensed data, especially to collect information on many samples quickly. However, cluster sampling must be used intelligently. Simply using very large clusters is not a valid method of collecting data, because each pixel is not independent of the other and adds very little information to the cluster. Congalton (1988b) recommended that no clusters larger than 10 pixels and certainly not larger than 25 pixels be used because of the lack of information added by each pixel beyond these cluster sizes.

Finally, stratified systematic unaligned sampling attempts to combine the advantages of randomness and stratification with the ease of a systematic sample without falling into the pitfalls of periodicity common to systematic sampling. This method is a combined approach that introduces more randomness than just a random start within each strata.

Congalton (1988b) performed sampling simulations on three spatially diverse areas (see Figure 3-1) using all five of these sampling schemes and concluded that in all cases simple random and stratified random sampling provided satisfactory results.

Simple random sampling allows reference data to be collected simultaneously for both training and assessment. However, it is not always appropriate, because it tends to undersample rarely occurring, but possibly very important, map categories unless the sample size is significantly increased. For this reason, stratified random sampling, where a minimum number of samples are selected from each strata (i.e., map category), is often recommended. However, stratified random sampling can be impractical, because stratified random samples can only be selected after the map has been completed (i.e., when the location of the strata are known). This limits the accuracy assessment data to being collected late in the project instead of in conjunction with the training data collection, thereby increasing the costs of the project. In addition, in some projects the time between the project beginning and the accuracy assessment may be so long as to cause temporal problems in collecting ground reference data. In other words, the ground may change (e.g., the crop harvested) between the time the project is started and the accuracy assessment is begun.

The concept of randomness is a central issue when performing almost any statistical analysis because a random sample is one in which each member of the population has an equal and independent chance of being selected. Therefore, a random sample ensures that the samples will be chosen without bias. If photo or video interpretation is used to label reference samples, then random sampling is feasible because access to the samples will not be a problem. However, a subset of the sample should be visited on the ground to assess the accuracy of the interpretation.

Despite the nice statistical properties of random sampling, access in the field to random sample units can often be problematical because many of the samples will be difficult to locate. Locked gates, fences, travel distances, and rugged terrain all combine to make random field sampling extremely costly and difficult. In forested and other wildland environments, randomly selected samples may be totally inaccessible except by helicopter. The cost of getting to each of the randomly located samples can be more than the cost of the rest of the entire mapping effort.

Obviously, one cannot spend the majority of a project's resources collecting accuracy assessment reference data. Instead, some balance must be struck. Often some combination of random and systematic sampling provides the best balance between statistical validity and practical application. Such a system may employ systematic or simple random sampling to collect some assessment data early in a project, and stratified random sampling within strata after the classification is completed to assure enough samples were collected for each category and to minimize any periodicity in the data. However, results of Congalton (1988a) showed that periodicity in the errors as measured by the autocorrelation analysis could make the use of systematic sampling risky for accuracy assessment.

An example of a combined approach could include a systematic sample tied to existing aerial photography with sample selection based on the center of every nth photo. Sample choices based on flight lines should not be highly correlated with a factor determining land cover unless the flight lines were aligned with a landscape feature. Choice of the number of samples per photo and the sampling interval between photos would depend on the size of the area to be mapped and the number of samples to collect. This systematic sample would assure that the entire mapped area gets covered.

However, rarely occurring map classes will probably be undersampled. It may be necessary to combine this approach with a stratified random sample when the map is completed to augment the underrepresented map categories. It may be practical to bound the stratified random field sample selection within some realistic distance of the roads. However, care must be taken, because roads tend to occur on flatter areas and in valleys along streams, which will bias sample selection to land cover likely to exist there. Care needs to be taken to mitigate these factors so that the most representative sample can be achieved. This type of combined approach minimizes the resources used and obtains the maximum information possible. Still, the statistical complexities of such a combination cannot be neglected. Again, a balance is desirable.

Finally, some analytic techniques assume that certain sampling schemes were used to obtain the data. For example, use of the Kappa analysis for comparing error matrices (see Chapter 5 for details on this analysis technique) assumes a multinomial sampling model. Only simple random sampling completely satisfies this assumption. If another sampling scheme or combination of sampling schemes is used, then it may be necessary to compute the appropriate variance equations for performing the Kappa analysis or other similar techniques. The effects (i.e., bias) of using another of the sampling schemes discussed here and not computing the appropriate variances is unknown.

An interesting project would be to test the effect on the Kappa analysis of using a sampling scheme other than simple random sampling. If the effect is found to be small then the scheme may be appropriate to use within the conditions discussed above. If the effect is found to be large, then that sampling scheme should not be used to perform Kappa analysis. If that scheme is to be used then the appropriate correction to the variance equation must be applied. Stehman (1992) has done such an analysis for two sampling schemes (simple random sampling and systematic sampling). His analysis shows that the effect on the Kappa analysis of using systematic sampling is negligible. This result adds further credence to the idea of using a combined systematic initial sample followed by a random sample to fill in the gaps.

Because of the many assumptions required for statistical analysis, a few researchers have concluded that some sampling schemes can be used for descriptive techniques and others for analytical techniques. However, this conclusion seems quite impractical. Accuracy assessment is expensive and no one is going to collect data for only descriptive use. Eventually, someone will use that matrix for some analytical technique. It is best to pay close attention to both the practical limitations and the statistical requirements when performing any accuracy assessment.

CHAPTER 4

Data Collection

Accuracy assessment data collection requires completing three steps using both the reference data and the map being assessed:

- First, the accuracy assessment sample site must be located both on the reference data and on the map. This can be a relatively simple task in an urban area, or far more difficult in a wildland where few recognizable landmarks exist.
- Next, the sample unit must be delineated. Sample units must be exactly the same area on both the reference data and the map. Usually they are delineated once on either the reference data or map.
- Finally, reference and map data must be collected for each sample unit to create reference and map labels based on the map classification scheme. The reference data may be collected from a variety of sources, and may be captured either through observation or measurement.

Serious oversights and problems can arise at each step of data collection. To adequately assess the accuracy of the remotely sensed classifications, each step must be implemented correctly on every sample. If the reference data is inaccurate, then the entire assessment becomes meaningless. Four basic considerations drive all reference data collection:

1. What should be the source data for the reference samples? Can existing maps or existing field data be used as reference data? Should the information be collected from aerial sources or field visits?
2. What type of information should be collected for each sample? Should measurements be taken or are observations adequate?
3. When should the reference data be collected? During initial field investigations when the map is being made, or only after the map is completed? What are the implications of using old data for accuracy assessment?
4. How do we insure that the reference data is collected correctly, objectively, and consistently?

There are many methods for collecting reference data, some of which depend on making observations (qualitative assessments) and some which require detailed, quantitative measurements. Given the varied reliability, difficulty, and expense of collecting this information, it is critical to know which of these data collection techniques are valid and which are not for any given project.

WHAT SHOULD BE THE SOURCE OF THE REFERENCE DATA?

The first decision in data collection requires determining what source data will be used for the determination of reference labels. Maps are rarely 100% correct. Each remote sensing project requires trade-offs between the remotely sensed data used to create the map and the level of accuracy required by the project. We accept some level of error as a trade-off for the cost savings inherent in using remotely sensed data.

However, accuracy assessment reference data must be 100% correct if it is to be a fair assessment of the map. Thus, reference labels must be collected from source data that is assumed to be more reliable than the remotely sensed data used to make the map. The type of source data required will depend upon the complexity of the map classification scheme. As a general rule, the simpler the classification scheme, the simpler the reference data collection.

Sometimes previously existing maps or ground data are used. Usually the source data are newly collected information that is one step closer to the ground than the remotely sensed data used to make the map. Thus, aerial photography is often used to assess the accuracy of maps made from satellite imagery, and ground visits are often used to assess the accuracy of maps created from aerial photography.

Using Existing versus Newly Collected Data

When a new map is produced, usually the first reaction is to compare the map to some existing source of information. Using previously collected ground information or existing maps for accuracy assessment is tempting because of the cost savings resulting from avoiding new data collection. While this can be a valuable qualitative tool, existing data is rarely acceptable for accuracy assessment because:

1. The classification systems employed in existing information usually differ from the one being used to create the new map. Comparisons between the two maps can result in the error matrix including differences between the reference data and the map data that do not measure map error, but are caused solely by differences in classification systems.
2. Existing data are older than those being used to create the new map. Changes on the landscape will not be reflected in the existing data. However, differences in the error matrix caused by the changes will be incorrectly assumed to be caused by map error.
3. Errors in the existing data are rarely known. Usually the errors are blamed on the new classification, thereby wrongly lowering the new map's accuracy. It is exactly this problem that has been one of the main reasons for the lack of acceptance of digital satellite data for many applications.

If existing information is the only available source of reference data, then consideration should be given to not performing quantitative accuracy assessment. Instead, a qualitative comparison of the new map and existing information should be performed, and the differences between the two should be analyzed.

Photos versus Ground

If new data is to be collected for reference samples, then a choice must be made between using ground visits versus aerial photography, video, or reconnaissance as the source reference data. The accuracy assessment professional must assess the reliability of each data type to obtain a correct label for the reference sample site.

Simple classification schemes with a few general classes can often be reliably assessed from air reconnaissance or interpretation of aerial photography or video. As the level of detail in the map classification scheme increases, so does the complexity of the reference data collection. Eventually, even very large scale photography cannot provide valid reference data. Instead, the data must be collected on the ground.

In some situations, the use of photo interpretation or videography for generating reference data may not be appropriate. For example, aerial photo interpretation is often used as reference data for assessing a land cover map generated from satellite imagery such as Landsat Thematic Mapper. The photo interpretation is assumed correct because it has greater spatial resolution than the satellite imagery and because photo interpretation has become a time-honored skill that is accepted as accurate. Unfortunately, errors also occur in photo interpretation and air reconnaissance depending on the skill of the photo interpreter and the level of detail required by the classification system. Inappropriately using photo interpretation as reference data could severely bias the conclusions about the accuracy of the satellite-based land cover map. In other words, one may conclude that the satellite-based map is of poor accuracy when actually it is the reference data that is inaccurate.

In such situations, actual ground visitation may be the only reliable method of data collection. At the very least, a subset of data should be collected on the ground and compared with the airborne data to verify the reliability of the airborne reference data. Even if the majority of reference data will come from photo interpretation or videography, it is critical that a subsample of these areas get visited in the field to verify the reliability of the interpretation. Much work is yet to be done to determine the proper level of effort and collection techniques necessary to provide this vital information. When the agreement between the interpretation and the ground begins to disagree regularly, it is time to switch to ground-based reference data collection. However, the collection of ground reference data is extremely expensive, and therefore the collection effort must be sufficient to meet the needs of the accuracy assessment while being efficient enough to meet the needs of the budget.

In a pilot study, Biging et al. (1991) compared photo interpretation to ground measurements for characterizing forest structure. These characteristics included forest species, tree size class, and crown closure. The ground data used for comparison were a series of measurements made in a sufficient number of ground plots to characterize each forest polygon (i.e., stand). The results showed that photo interpretation of species ranged in accuracy between 75% and 85%. The accuracy of size class was around 75% and the accuracy of crown closure was less than 40%. This study reinforces the need to be careful if assuming that the results of the photo interpretation are sufficient for use as reference data in an accuracy assessment.

HOW SHOULD THE REFERENCE DATA BE COLLECTED?

The next decision involves deciding how information will be collected from the source data to obtain a reliable label for each reference site. Reference data must be collected using the same classification scheme that was used for the remotely sensed data, and should also be applied over the same minimum mapping unit as was applied to the remotely sensed data. In many instances, simple observations/interpretations are sufficient for labeling a reference sample. In other cases, observation is not enough and actual measurements in the field are required.

The purpose of collecting reference data for a sample site is to derive a "true" label for the site for comparison to the map label. Often the reference label can be obtained by merely observing the site from an airplane, car, or aerial photography. For example, in most cases a golf course can be accurately identified through observation.

Whether or not accuracy assessment reference data should be obtained from observations or measurements will be determined by the complexity of the landscape, the detail of the classification system, the required precision of the accuracy assessment, and the project budget. Reference data for simple classification schemes that distinguish homogeneous land cover types from one another usually can be obtained from observations and/or estimations either on the ground or from other remotely sensed data such as aerial photography. For example, distinguishing conifer forest from an agricultural field from a golf course can be determined from observation. Collecting reference data may be as simple as looking at aerial photography or observing sites on the ground.

However, complex classification systems may require measurement to determine precise (i.e., nonvarying) reference site labels. For example, a more complex forest classification scheme may involve collecting reference data for tree size class (diameter of the trunk). Tree size class is important both as a determinant of spotted owl habitat and as a measurement of wood products merchantability. Size class can be occularly estimated in photos and on the ground. However, different individuals may make different estimations introducing variability into the observation. Not only will this variation exist between individuals, but also within one individual. The same observer may see things differently depending on whether it is Monday or Friday; or whether it is sunny or raining; or especially depending on how much coffee he or she has consumed. To avoid the variability, size class can be measured, but a great many trees will need to be measured to estimate the size class for each sample unit. In such instances the accuracy assessment professional must decide whether the project requires measurement (which can be time-consuming and expensive) or if the variation inherent in observed estimations can be accepted.

Whether or not measurements are required depends on the level of precision required by the map users and on the project budget. Information on spotted owl habitat requirements indicate that the owls prefer older multistoried stands that include large trees. "Large" in this context is relative, and precise measurements of trees will probably not be needed as long as the map accurately distinguishes between stands of single storied small trees and multistoried large trees. In contrast, many wood products mills can only accept trees within a specific size class. Trees one inch smaller or larger than the prescribed range cannot be accepted by the machinery in the mill. In this case, measurement will probably be required.

Observer variability is especially evident in estimates of vegetation cover, which cannot be precisely measured from aerial photographs. In addition, ground verification of aerial estimates of vegetative cover is problematic, as estimates of cover from the ground (i.e., below tree canopies) are fundamentally different from estimates made from above the canopy. Estimates from below may include vegetative cover from small shrubs and trees that cannot be seen from above by the remotely sensed data, be it from aerial photography or satellite imagery. Therefore, using ground estimates as reference data for aerial cover estimates can be like comparing "apples and oranges."

The trade-offs inherent between observation and measurement are exemplified in a pilot study conducted to determine the level of effort needed to collect appropriate ground reference data for use in forest inventory. The objective of this study was to determine if visual calls made by trained experts walking into forest polygons are sufficient or whether actual ground measurements need to be made. There are obviously many factors influencing the accuracy of ground data collection, including the complexity of the vegetation itself. A variety of vegetation complexities were represented in this study. The results are enlightening to those remote sensing specialists who routinely collect forest ground data only by visual observation. The pilot study was part of a larger project aimed at developing the use of digital remotely sensed data for commercial forest inventory (Biging and Congalton 1989).

Commercial forest inventory involves much more than creating a land cover map derived from digital remotely sensed data. Usually the map is used only to stratify the landscape; a field inventory is conducted on the ground to determine tree volume statistics for each type of stand of trees. A complete inventory requires that the forest type, size class, and crown closure of a forested area be known in order to determine the volume of the timber in that area. If a single species dominates, the forest type is commonly named by that species (Eyre 1980). However, if a combination of species are present, then a mixed label is used (e.g., the mixed conifer type). The size of the tree is measured by the diameter of the tree at 4.5 feet above the ground (i.e., diameter at breast height, DBH) and then is divided into size classes such as poles, small saw timber, and large saw timber. This measure is obviously important, because large diameter trees contain more volume (i.e., valuable timber) than small diameter trees. Crown closure as measured by the amount of ground area the tree crowns occupy (canopy closure) is also an important measure of tree size and numbers. Therefore, in this pilot study, it was necessary to collect ground reference data not only on tree species/type, but on crown closure and size class as well.

Ground reference data were collected using two approaches. In the first approach, a field crew of four entered a forest stand (i.e., polygon), observed the vegetation, and came to a consensus for a visual call of dominant species/type, size class of the dominant species, crown closure of the dominant size class, and crown closure of all tree species combined. Dominance was defined as the species or type comprising the majority of forest volume. In the second approach, measurements were conducted on a fixed-radius plot to record the species, DBH, and height of each tree falling within the plot. A minimum of two plots (1/10 or 1/20 acre) were measured for each forested polygon. Because of the difficulty of making all the required measurements (precise location and crown width for each tree in the plot) to estimate crown closure on the plot, an approach using transects was developed to determine crown closure. A minimum of four 100 foot long transects randomly located within the polygon were used to collect crown closure information. The percent of crown closure was determined by the presence or absence of tree crown at 1 foot intervals along the transects. All the measurements were input into a computer program that summarized the results into the dominant species/type, the size class of the dominant species/type, the crown closure of the dominant size class, and the crown closure of all tree species for each forested area. The results of the two approaches were compared by using an error matrix.

Table 4-1 shows the results of field measurement versus visual call as expressed in an error matrix for the dominant species. This table indicates that species can be fairly well determined from a visual call because there is strong agreement between the field measurements and the visual call. Of course, this conclusion requires one to assume that the field measurements are a better measure of ground reference data, a reasonable assumption in this case. Therefore, ground reference data collection of species information can be maximized using visual calls, and field measurements appear to be unnecessary.

Table 4-2 presents the results of comparing the two ground reference data collection approaches for the dominant size class. As in species, the overall agreement is relatively high with most of the confusion occurring between the larger classes. The greatest inaccuracies result from visually classifying the dominant size class (i.e., the one with the most volume) as size class three (12–24 inch DBH) when in fact size class four (>24 inch DBH) trees contained the most volume. This visual classification error is easy to understand. Tree volume is directly related to the square of DBH. There are numerous cases when a small number of large trees contribute the majority of the volume in the stand, while there may be many more medium size trees present. The dichotomy between prevalence of medium size trees but dominance in volume by a small number of trees can be difficult to assess visually. It is likely that researchers and practitioners would confuse these classes in cases where the size class with the majority of volume was not readily evident. In cases like this, simply improving one's ability to visually estimate diameter would not improve one's ability to classify size class. The ability to weigh numbers and sizes to estimate volume requires considerable experience and would certainly require making plot and tree measurements to gain and retain this ability.

Tables 4-3 and 4-4 show the results of comparing the two collection approaches for crown closure. Table 4-3 presents the crown closure of the dominant size class results, while Table 4-4 shows the results of overall crown closure. In both matrices, there is very low agreement (46–49%) between the observed estimate and the field measurements.

Table 4-1   **Error Matrix for the Field Measurement versus Visual Call for Dominant Species. Reproduced with permission, the American Society for Photogrammetry and Remote Sensing, from: Congalton, R. and G. Biging, 1992. A pilot study evaluating ground reference data collection efforts for use in forest inventory.** *Photogrammetric Engineering and Remote Sensing.* **Vol. 58, No. 12, pp. 1669-1671.**

VisualCall	Field Measurement								
		TF	MC	LP	DF	PP	PD	OAK	row total
	TF	14	0	0	0	0	0	0	14
	MC	0	10	0	0	0	2	0	12
	LP	0	0	1	0	0	0	0	1
	DF	0	1	0	8	0	0	0	9
	PP	1	1	0	0	0	0	0	2
	PD	0	0	0	1	0	0	0	1
	OAK	0	0	0	0	0	0	0	0
	column total	15	12	1	9	0	2	0	39

Species

TF = true fir
MC = mixed conifer
LP = lodgepole pine
DF = Douglas fir
PP = Ponderosa pine
PD = PP and DF
OAK = oaks

OVERALL ACCURACY = 33/ 39 = 85%

PRODUCER'S ACCURACY

TF = 14/ 15 = 93%
MC = 10/12 = 83%
LP = 1/ 1 = 100%
DF =   8/9 = 89%
PP = 0/ 0 = —
PD = 0/ 2 = 0%
OAK = 0/0   = —

USER'S ACCURACY

TF = 14/14 = 100%
MC = 10/12 = 83%
LP =   1/1 = 100%
DF =   8/9 = 89%
PP = 0/2 =   0%
PD = 0/1 =   0%
OAK =   0/ 0 = —

Therefore, it appears that field measurements must be used to obtain precise measures of crown closure and that visual calls, although less expensive and quicker, may vary at an unacceptable level.

In conclusion, it must be emphasized that this is only a small pilot study. Further work needs to be conducted in this area to evaluate ground reference data collection methods and to include the validation of aerial methods (i.e., photo interpretation and videography). The results presented demonstrate that making visual calls of species are relatively easy and accurate, except where many species occur simultaneously. Size class is more difficult to assess than species, because of the implicit need to estimate the size class with the majority of volume. Crown closure is by far the toughest to determine. It is most dependent on where one is standing when the call is made. Field measurements, such as the transects used in this study, provide a better means of determining crown closure. This study has shown that at least some ground data must be collected using measurements, and it has suggested that a multilevel effort may result in the most efficient and practical method for collection of ground reference data.

Table 4-2 Error Matrix for the Field Measurement versus Visual Call for Dominant Size Class. Reproduced with permission, the American Society for Photogrammetry and Remote Sensing, from: Congalton, R. and G. Biging, 1992. A pilot study evaluating ground reference data collection efforts for use in forest inventory. *Photogrammetric Engineering and Remote Sensing.* Vol. 58, No. 12, pp. 1669-1671.

Field Measurement

VisualCall	1	2	3	4	row total
1	1	0	0	0	1
2	1	3	1	0	5
3	0	0	17	5	22
4	0	0	1	11	12
column total	2	3	19	16	40

Size Classes

1 = 0-5" dbh

2 = 5-12" dbh

3 = 12-24" dbh

4 = >24" dbh

OVERALL ACCURACY
= 32/40 = 80%

PRODUCER'S ACCURACY					USER'S ACCURACY			
1	=	1/2	=	50%	1	=	1/1	= 100%
2	=	3/3	=	100%	2	=	3/5	= 60%
3	=	17/19	=	89%	3	=	17/22	= 77%
4	=	11/16	=	69%	4	=	11/12	= 92%

Table 4-3 Error Matrix for the Field Measurement versus Visual Call for Density (Crown Closure) of the Dominant Species. Reproduced with permission, the American Society for Photogrammetry and Remote Sensing, from: Congalton, R. and G. Biging, 1992. A pilot study evaluating ground reference data collection efforts for use in forest inventory. *Photogrammetric Engineering and Remote Sensing.* Vol. 58, No. 12, pp. 1669-1671.

Field Measurement

VisualCall	O	L	M	D	row total
O	10	8	3	0	21
L	2	8	1	1	12
M	0	3	1	1	5
D	0	1	0	0	1
column total	12	20	5	2	39

Density Classes

O = Open

L = Low

M = Medium

D = Dense

OVERALL ACCURACY
= 19/39 = 49%

PRODUCER'S ACCURACY					USER'S ACCURACY			
O	=	10/12	=	83%	O	=	10/21	= 48%
L	=	8/20	=	40%	L	=	8/12	= 67%
M	=	1/5	=	20%	M	=	1/5	= 20%
D	=	0/2	=	0%	D	=	0/1	= 0%

Table 4-4   Error Matrix for the Field Measurement versus Visual Call for Overall Density (Crown Closure). Reproduced with permission, the American Society for Photogrammetry and Remote Sensing, from: Congalton, R. and G. Biging, 1992. A pilot study evaluating ground reference data collection efforts for use in forest inventory. *Photogrammetric Engineering and Remote Sensing.* Vol. 58, No. 12, pp. 1669-1671.

Field Measurement

Visual Call	O	L	M	D	row total	
O	0	1	1	0	2	
L	1	3	7	0	11	
M	0	0	8	10	18	
D	0	0	0	6	6	
column total	1	4	16	16	37	

Density Classes

O = Open

L = Low

M = Medium

D = Dense

OVERALL ACCURACY
= 17/ 37 = 46%

PRODUCER'S ACCURACY

O =   0/1   = 0%
L =   3/ 4   = 75%
M =   8/ 16 = 50%
D =   6/ 16 = 38%

USER'S ACCURACY

O =   0/2   = 0%
L =   3/11   = 27%
M =   8/ 18 = 44%
D =   6/6   = 100%

WHEN TO COLLECT REFERENCE DATA

The world's landscape is constantly changing. If change occurs between the date of capture of the remotely sensed data used to create a map and the date of reference data collection, accuracy assessment reference sample labels may be affected. When a crop is harvested, a wetland drained, or a field developed into a shopping mall, the error matrix may show a difference between the map and the reference label that is not caused by map error, but rather by landscape change.

For example, aerial photography is often used as reference source data for accuracy assessment of forest type maps created from Landsat TM or SPOT satellite data. Because aerial photography is relatively expensive to obtain, existing photography often 5 to 15 years old is used. If an area has changed because of fire, disease, harvesting, or growth, the resulting reference labels in the changed areas will be incorrect. Harvests and fire are clearly visible on most satellite imagery, making it possible to detect the changes by looking at the imagery.* However, stand growth and partial defoliation from disease are not as readily observable on the imagery,

* Using the satellite imagery to correct the reference information collected from the photos seems a little convoluted since the photos are supposedly being used to assess the accuracy of a map produced from the imagery.

making the use of older photos especially problematic in the Northwest and Southeast, where trees can grow through several size classes in a 10-year period.

Therefore, accuracy assessment reference data should be collected as close as possible to the date of the collection of the remotely sensed data used to make the map. However, trade-offs may need to be made between the timeliness of the data collection and the need to use the resulting map to stratify the accuracy assessment sample. In most remote sensing mapping projects it is necessary to go to the field to get familiar with the area to be mapped and to collect information for training the classifier (i.e., supervised classification) or to aid in labeling the clusters (i.e., unsupervised classification). If reference data for accuracy assessment can be collected independently, but simultaneously, then a second trip to the field is eliminated, saving costs and ensuring that reference data collection is occurring close to the time the remotely sensed data is captured.

However, if accuracy assessment reference data are collected at the beginning of the project before the map is generated, then it is not possible to stratify by map class since the map has yet to be created. It is also not possible to have a proportional to area allocation of the sample since the total area of each map class is still unknown.

For example, the U.S.D.I. Bureau of Reclamation maps the crops of the Lower Colorado River Region four times a year using Landsat TM data. Farm land in this region is so productive and valuable that growers plant three to four crops per year and will plow under a crop to plant a new one in response to the future's market. With so much crop change, ground data collection and accuracy assessment must occur at the same time the imagery is collected. The Bureau fields a ground data collection crew for 2 weeks surrounding the date of image acquisition. A random number generator is used to determine the fields to be visited and the same fields are visited during each field effort, regardless of the crops being grown. Therefore, the accuracy assessment sample is random, but not stratified by crop type. As Table 4-5 illustrates, some crops are oversampled and others are undersampled each time. The Bureau believes it is more important to ensure correct crop identification than it is to ensure that enough samples are collected in rarely occurring crop types.

ENSURING OBJECTIVITY AND CONSISTENCY

For accuracy assessment to be useful, map users must have faith that the assessment is an exact representation of the map's accuracy. They must believe that the assessment is objective and the results are repeatable. Maintaining the following three conditions will ensure objectivity and consistency:

1. Accuracy reference data must always be kept independent of any training data.
2. Data must be collected consistently from sample site to sample site.
3. Quality control procedures must be developed and implemented for all steps of data collection.

Table 4-5   Error Matrix Showing Number of Samples in Each Crop Type

REFERENCE DATA

	A	C	SG	CN	L	M	BG	CS	T	SU	O	CR	F	D	S	Total
A	157		8				3						3			171
C		1		1	1											3
SG	3		163	6							12	2	1			187
CN																0
L			4	3						1		1				9
M					5							1				6
BG	1						10									11
CS							69									69
T																0
SU																0
O			1	3							7					11
CR												2				2
F													224			224
D														11		11
S																0
Total	161	1	176	0	13	6	13	69	0	1	19	5	229	11	0	704

(Left axis: MAP DATA)

LEGEND			Producer's Accuracy	User's Accuracy
A	= Alfalfa	A	98%	92%
C	= Cotton	C	100%	33%
SG	= Small Grains	SG	93%	87%
CN	= Corn	CN	---	---
L	= Lettuce	L	23%	33%
M	= Melons	M	83%	83%
BG	= Bermuda Grass	BG	77%	91%
CS	= Citrus	CS	100%	100%
T	= Tomatoes	T	---	---
SU	= Sudan Grass	SU	0%	---
O	= Other Veg.	O	37%	64%
CR	= Crucifers	CR	40%	100%
F	= Fallow	F	98%	100%
D	= Dates	D	100%	100%
S	= Safflowers	S	---	---

Data Independence

It was not uncommon for early accuracy assessments to use the same information to assess the accuracy of a map as was used to create the map. This unacceptable procedure obviously violates all assumptions of independence and biases the assessment in favor of the map. Independence of the reference data can be assured in one of two ways. First, the reference and training data collection can be performed at a completely different time and/or by different people. Collecting information at different times is both expensive and can introduce landscape change problems, as discussed above. Using different people can also be expensive, as more personnel need to be completely trained in the details of the project.

The second method for ensuring independence involves collecting reference and training data simultaneously and then using a random number generator to select and remove the accuracy assessment sites from the training data set. The accuracy assessment sites are not reviewed again until it is time to perform the assessment. In both cases, accuracy assessment reference data must be kept absolutely independent (i.e., separate) from any training/labeling data.

Data Collection Consistency

Data collection consistency can be assured through personnel training and the development of objective data collection procedures. Training should occur simultaneously for all personnel at the initiation of data collection. One to three days of intensive training is often necessary and must include reference collection on numerous example sites that represent the broad array of variation both between and within map classes. Trainers must insure that reference data collection personnel are (1) applying the classification scheme correctly and (2) accurately identifying characteristics of the landscape that are inherent in the classification scheme. For example, if a classification scheme depends on the identification of plant species, then all reference data personnel must be able to accurately identify species on the reference source data. The classification scheme must also be applied over the same minimum mapping unit as was applied to create the map.

In addition to personnel training, objective data collection procedures are key to consistent data collection. The more measurement (as opposed to estimation) involved in reference data collection, the more consistent and objective the collection. However, measurement increases the cost of accuracy assessment, so most assessments rely heavily on ocular estimation. If ocular estimates are to be used, then the variance inherent in estimation must be accepted as an unavoidable part of the assessment, and some method of assessing the variance must be included in the assessment. Several of these methods are discussed in Chapter 7, Advanced Topics.

An important mechanism for imposing objectivity is the use of a reference data collection form to force all data collection personnel through the same collection process. The complexity of the reference data collection form will depend on the level of the complexity of the classification scheme. The form should lead the collector through a quantitative process to a definitive answer from the classification scheme. It also provides a means of performing a quality assessment/quality control check on the collection process. Figure 4-1 is an example data collection form for a relatively simple classification scheme. An important portion of this form is the dichotomous key that leads data collection personnel to the land cover class label based solely on the classification scheme rules.

Reference data collection forms, regardless of their complexity, have some common components. These include (1) the name of the collector and the date of the

Site Name: _____   Site ID #: _____

Quad name: _____   Satellite Scene ID #: _____

Latitude: _____ Longitude: _____ Elevation range: _____

Method for determining position: _____

Comments on Position: _____

Comments on Weather: _____

Crew names: _____   Date: _____

CATEGORIES

Background	Forest	Palustrine forest
Developed	Scrub/Shrub	Estuarine emergent
Crop/Grass	Exposed Land	Palustrine emergent
	Water	

Sample Unit

Vegetated — **Non-vegetated**

Water present

If tree cover > 10%
PALUSTRINE FOREST

If tree cover < 10% &
fresh water
PALUSTRINE EMERGENT

If tree cover < 10% &
salt water
ESTUARINE EMERGENT

Water absent

If tree cover > 10%
FOREST

If tree cover < 10% &
other woody veg > 25%
SCRUB/SHRUB

If tree cover < 10% &
other woody veg < 25%
CROP/GRASS

If person-made structures
DEVELOPED

If no structures
EXPOSED LAND

WATER

If none of the above
BACKGROUND

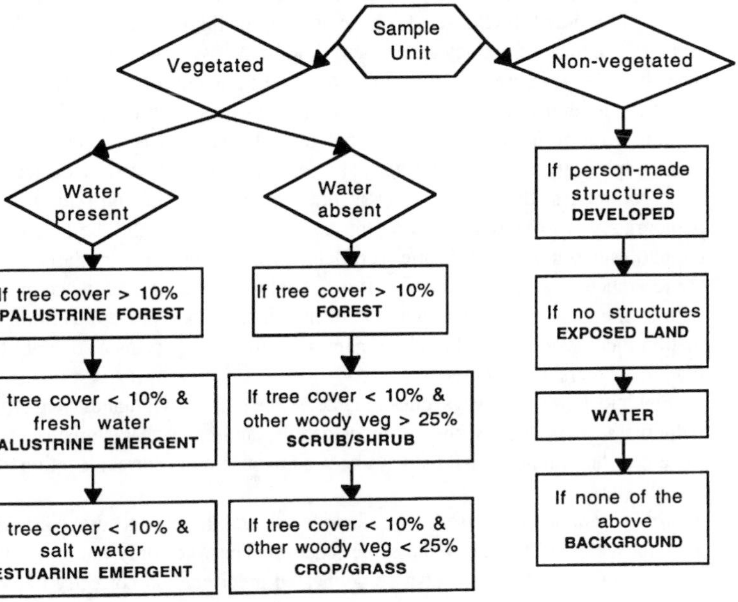

Actual Category as determined from flow chart: _____

Comments on anomalies, variability, or interesting findings: _____

Figure 4-1 Example of a reference data collection form for a simple classification scheme.

collection, (2) locational information about the site, (3) some type of table or logical progression that represents what the collector is seeing, (4) a place to fill in the actual category name from the classification scheme, and (5) a place to describe any anomalies, any variability, or interesting findings at the site.

Quality Control

Quality control is necessary at every step of data collection. Each error in data collection can translate into an incorrect indication of map accuracy. Data collection errors result in both over- and under-estimations of map accuracy.

The following text discusses some of the most common quality control problems in each step of accuracy assessment data collection. Because accuracy assessment requires collecting information from both the reference source data and the map, each step involves two possible occasions of error: during collection from the map, and during collection from the reference source data.

1. *Location of the accuracy assessment sample site.* It is not uncommon for accuracy assessment personnel to collect information on the wrong location because inadequate procedures were used to locate the site on either the map or the reference data. As discussed in Chapter 3, either the map or the reference source data can be used to allocate accuracy assessment samples across space. If the map is used, then the accuracy assessment samples must be transferred to the reference data. When the reference data is used, then the samples must be transferred to the map. In either case, if the reference data is not geocoded (e.g., as is the case with aerial photography), then accurate location and transfer of the site can be difficult.

   A common method for locating accuracy assessment sites on reference aerial photography is to view the site on the map and then "eyeball" the location onto the photos based on similar patterns of land cover and terrain in both the map and the reference data. In this situation, it is critical to provide the reference personnel with as much information as possible to help them locate the site. Helpful information includes digitized flightline maps and other ancillary data such as stream, road, or ownership coverages.

   Field location is always questionable, especially in wildlands (e.g., tundra, wilderness areas, etc.) with few recognizable landscape characteristics. GPS is extremely helpful and should always be used to ensure the correct location of field sample sites.

2. *Sample unit delineation.* Both the reference site and the map accuracy assessment sites must represent exactly the same location. Thus, not only must the sites be properly located, they must also be delineated precisely and correctly transferred to a planimetric base. For example, if an existing map is used as the reference source, and the map was not registered correctly, then all accuracy assessment reference sites will not register to the new map being assessed and a misalignment will occur when the reference site and the map site are compared. This is not an uncommon situation when aerial photography was used to create the existing map, and the transfer from the photo to the map was performed ocularly without the use of photogrammetric equipment.

   Another common error in accuracy assessment occurs when the reference and the map sites are in the same general location, but are of different sizes or shapes. For example, if map polygons constitute the population to be sampled, and the

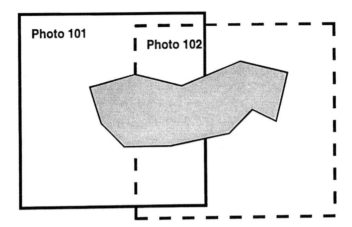

Figure 4-2 The shaded accuracy assessment site polygon falls across two aerial photographs.

reference data is to be aerial photography, then the selected polygons will often fall across two or more photos, as depicted in Figure 4-2. In this case, the analyst must either collect reference data across all the photos (which can be time consuming), or the sample site must be reduced in size on both the map and the reference data so that it fits on one photo.

3. Data collection and data entry are the most common sources of quality control problems in accuracy assessment. Data collection errors occur when measurements are done incorrectly, variables of the classification scheme are misidentified (e.g., species), or the classification scheme is misapplied. In addition, weak classification schemes will also create ambiguity in data collection. Unfortunately, the first indication of a weak classification system often occurs during accuracy assessment, when the map is already completed, and refinement of the classification scheme is not possible.

   Data collection errors are usually monitored by selecting a subsample of the accuracy assessment sites and having reference data collected simultaneously on them by two different personnel. Usually, the most experienced personnel are assigned to the subsample. When differences are detected, the source of the differences are immediately identified. If data collection errors are the source of the differences, then personnel are either retrained or removed from reference data collection.

   When aerial photography is used as the reference source data, it is critical that a ground assessment of the photo interpretation be conducted. In addition, reference samples chosen from an existing map must also be assessed for accuracy.

   Data entry errors can be reduced by using digital data entry forms that restrict each field of the form to an allowable set of characters. Data can also be entered twice and the two data sets compared to identify differences and errors. Data entry errors also can occur when the site is digitized. Quality control must include a same scale comparison of the digitized site to the source map.

Finally, although no reference data set may be completely accurate, it is important that the reference data have high accuracy or else it is not a fair assessment. Therefore, it is critical that reference data collection be carefully considered in any accuracy assessment. Much work is yet to be done to determine the proper level of effort and collection techniques necessary to provide this vital information.

CHAPTER **5**

Basic Analysis Techniques

This chapter presents the basic analysis techniques needed to perform an accuracy assessment. The chapter begins by discussing early non-site specific assessments. Next, site specific assessment techniques employing the error matrix are presented followed by all the analytical tools that proceed from it including computing confidence intervals, testing for significant differences, and correcting area estimates. A numerical example is presented through the entire chapter to aid in understanding of the concepts.

NON-SITE SPECIFIC ASSESSMENTS

In a non-site specific accuracy assessment, only total areas for each category mapped are computed without regard to the location of these areas. In other words, a comparison between the number of acres or hectares of each category on the map generated from remotely sensed data and the reference data is performed. In this way, the errors of omission and commission tend to compensate for each other and the totals compare favorably. However, nothing is known about any specific location on the map or how it agrees or disagrees with the reference data.

A simple example quickly demonstrates the shortcomings of the non-site specific approach. Figure 5-1 shows the distribution of the forest category on both a reference image and two different classifications generated from remotely sensed data. Classification #1 was generated using one type of classification algorithm (e.g., supervised, unsupervised, or nonparametric, etc.) while classification #2 employed a different algorithm. In this example, only the forest category is being compared. The reference data shows a total of 2,435 acres of forest while classification #1 shows 2,322 acres and classification #2 shows 2,635 acres. In a non-site specific assessment, you would conclude that classification #1 is better for the forest category, because the total number of forest acres for classification #1 more closely agrees with the number of acres of forest on the reference image (2,435 acres – 2,322 acres = 113 acres difference for classification #1 while classification #2 differs by 200 acres).

Reference Data

total acres of forest = 2,435

Classified Image #1

total acres of forest = 2,322

Reference Data

total acres of forest = 2,435

Classified Image #2

total acres of forest = 2,635

Figure 5-1 Example of non-site specific accuracy assessment.

However, a visual comparison between the forest polygons on classification #1 and the reference data demonstrates little locational correspondence. Classification #2, despite being judged inferior by the non-site specific assessment, appears to agree in location much better with the reference data forest polygons. Therefore, the use of non-site specific accuracy assessment can be quite misleading. In the example shown here, the non-site specific assessment actually recommends the use of the inferior classification algorithm.

SITE SPECIFIC ASSESSMENTS

Given the obvious limitations of non-site specific accuracy assessment, there was a need to know how the map generated from the remotely sensed data compared to the reference data on a locational basis. Therefore, site specific assessments were instituted. Initially, a single value representing the accuracy of the entire classification (i.e., overall accuracy) was presented. This computation was performed by comparing a sample of locations on the map with the same locations on the reference data and keeping track of the number of times there was agreement.

An overall accuracy level of 85% was adopted as representing the cutoff between acceptable and unacceptable results. This standard was first described in Anderson et al. (1976) and seems to be almost universally accepted despite there being nothing magic or even especially significant about the 85% correct accuracy level. Obviously, the accuracy of a map depends on a great many factors, including the amount of effort, level of detail (i.e., classification scheme), and the variability of the categories to be mapped. In some applications an overall accuracy of 85% is more than sufficient and in other cases it would not be accurate enough. Soon after maps were evaluated on just an overall accuracy, the need to evaluate individual categories within the classification scheme was recognized, and so began the use of the error matrix to represent map accuracy.

The Error Matrix

As previously introduced, an error matrix is a square array of numbers set out in rows and columns that express the number of sample units (pixels, clusters, or polygons) assigned to a particular category in one classification relative to the number of sample units assigned to a particular category in another classification (Table 5-1). In most cases, one of the classifications is considered to be correct (i.e., reference data) and may be generated from aerial photography, airborne video, ground observation or ground measurement. The columns usually represent this reference data, while the rows indicate the classification generated from the remotely sensed data.

An error matrix is a very effective way to represent map accuracy in that the individual accuracies of each category are plainly described along with both the errors of inclusion (commission errors) and errors of exclusion (omission errors) present in the classification. A commission error is simply defined as including an area into a category when it does not belong to that category. An omission error is excluding that area from the category in which it truly does belong. Every error is an omission from the correct category and a commission to a wrong category.

For example, in the error matrix in Table 5-1 there are four areas that were classified as deciduous when the reference data show that they were actually conifer. Therefore, four areas were omitted from the correct coniferous category and committed to the incorrect deciduous category.

In addition to clearly showing errors of omission and commission, the error matrix can be used to compute other accuracy measures, such as overall accuracy, producer's accuracy, and user's accuracy (Story and Congalton 1986). Overall accuracy is simply the sum of the major diagonal (i.e., the correctly classified sample

Table 5-1   Example Error Matrix (same as presented in Table 2-1)

Reference Data

		D	C	AG	SB	row total
	D	65	4	22	24	115
	C	6	81	5	8	100
Classified Data	AG	0	11	85	19	115
	SB	4	7	3	90	104
column total		75	103	115	141	434

Land Cover Categories

D = deciduous

C = conifer

AG = agriculture

SB = shrub

OVERALL ACCURACY =
(65+81+85+90)/ 434 =
321/434 = 74%

PRODUCER'S ACCURACY	USER'S ACCURACY
D = 65/75 = 87%	D = 65/115 = 57%
C = 81/103 = 79%	C = 81/100 = 81%
AG = 85/115 = 74%	AG = 85/115 = 74%
SB = 90/141 = 64%	SB = 90/104 = 87%

units) divided by the total number of sample units in the entire error matrix. This value is the most commonly reported accuracy assessment statistic and is probably most familiar to the reader. However, just presenting the overall accuracy is not enough. It is important to present the entire matrix so that other accuracy measures can be computed as needed.

Producer's and user's accuracies are ways of representing individual category accuracies instead of just the overall classification accuracy. Before error matrices were the standard accuracy reporting mechanism, it was common to report the overall accuracy and either only the producer's or user's accuracy. A quick example will demonstrate the need to publish the entire matrix so that all three accuracy measures can be computed.

Studying the error matrix shown in Table 5-1 reveals an overall map accuracy of 74%. However, suppose we are most interested in the ability to classify hardwood forests, so we calculate a "producer's accuracy" for this category. This calculation is performed by dividing the total number of correct sample units in the deciduous category (i.e., 65) by the total number of deciduous sample units as indicated by the reference data (i.e., 75 or the column total). This division results in a "producer's accuracy" of 87%, which is quite good. If we stopped here, one might conclude that although this classification appears to be average overall, it is very adequate for the deciduous category. Making such a conclusion could be a very serious mistake. A quick calculation of the "user's accuracy" computed by dividing the total number of correct pixels in the deciduous category (i.e., 65) by the total number of pixels classified as deciduous (i.e., 115 or the row total) reveals a value of 57%. In other words, although 87% of the deciduous areas have been correctly

Table 5-2 Mathematical Example of an Error Matrix

		j = columns (reference)			row total n_{i+}
		1	2	k	
	1	n_{11}	n_{12}	n_{1k}	n_{1+}
i= rows (classification)	2	n_{21}	n_{22}	n_{2k}	n_{2+}
	k	n_{k1}	n_{k2}	n_{kk}	n_{k+}
column total n_{+j}		n_{+1}	n_{+2}	n_{+k}	n

identified as deciduous, only 57% of the areas called deciduous on the map are actually deciduous on the ground. A more careful look at the error matrix reveals that there is significant confusion in discriminating deciduous from agriculture and shrub. Therefore, although the producer of this map can claim that 87% of the time an area that was deciduous on the ground was identified as such on the map, a user of this map will find that only 57% of the time that the map says an area is deciduous will it actually be deciduous on the ground.

Mathematical Representation of the Error Matrix

This section presents the error matrix in mathematical terms necessary to perform the analysis techniques described in the rest of this chapter. The error matrix was presented previously in descriptive terms including an example (Table 5-1) that should help make this transition to equations and mathematical notation easier to understand.

Assume that n samples are distributed into k^2 cells where each sample is assigned to one of k categories in the remotely sensed classification (usually the rows) and, independently, to one of the same k categories in the reference data set (usually the columns). Let n_{ij} denote the number of samples classified into category i (i = 1, 2, ..., k) in the remotely sensed classification and category j (j = 1, 2, ..., k) in the reference data set (Table 5-2).

Let

$$n_{i+} = \sum_{j=1}^{k} n_{ij}$$

be the number of samples classified into category i in the remotely sensed classification, and

$$n_{+j} = \sum_{i=1}^{k} n_{ij}$$

be the number of samples classified into category j in the reference data set.

Overall accuracy between remotely sensed classification and the reference data can then be computed as follows:

$$\text{overall accuracy} = \frac{\sum_{i=1}^{k} n_{ii}}{n}.$$

Producer's accuracy can be computed by

$$\text{producer's accuracy}_j = \frac{n_{jj}}{n_{+j}}$$

and the user's accuracy can be computed by

$$\text{user's accuracy}_i = \frac{n_{ii}}{n_{i+}}.$$

Finally, let p_{ij} denote the proportion of samples in the i,jth cell, corresponding to n_{ij}. In other words, $p_{ij} = n_{ij}/n$.

Then let p_{i+} and p_{+j} be defined by

$$p_{i+} = \sum_{j=1}^{k} p_{ij}$$

and

$$p_{+j} = \sum_{i=1}^{k} p_{ij}.$$

Analysis Techniques

Once the error matrix has been represented in mathematical terms, then it is appropriate to document the following analysis techniques. These techniques clearly demonstrate why the error matrix is such a powerful tool and should be included in any published accuracy assessment. Without having the error matrix as a starting point, none of these analysis techniques would be possible.

Kappa

The Kappa analysis is a discrete multivariate technique used in accuracy assessment for statistically determining if one error matrix is significantly different than another (Bishop et al. 1975). The result of performing a Kappa analysis is a KHAT statistic (actually $\hat{K}$, an estimate of Kappa), which is another measure of agreement or accuracy (Cohen 1960). This measure of agreement is based on the difference between the actual agreement in the error matrix (i.e., the agreement between the remotely sensed classification and the reference data as indicated by the major diagonal) and the chance agreement which is indicated by the row and column totals (i.e., marginals). In this way the KHAT statistic is similar to the more familiar Chi square analysis.

Although this analysis technique has been in the sociology and psychology literature for many years, the method was not introduced to the remote sensing community until 1981 (Congalton 1981) and not published in a remote sensing journal before Congalton et al. (1983). Since then numerous papers have been published recommending this technique. Consequently, the Kappa analysis has become a standard component of most every accuracy assessment (Congalton et al. 1983, Rosenfield and Fitzpatrick-Lins 1986, Hudson and Ramm 1987, and Congalton 1991).

The following equations are used for computing the KHAT statistic and its variance. Let

$$p_o = \sum_{i=1}^{k} p_{ii}$$

be the actual agreement, and

$$p_c = \sum_{i=1}^{k} p_{i+} p_{+j} \qquad p_{i+} \text{ and } p_{+j} \text{ as previously defined above}$$

the "chance agreement."

Assuming a *multinomial sampling model*, the maximum likelihood estimate of Kappa is given by

$$\hat{K} = \frac{p_o - p_c}{1 - p_c} .$$

For computational purposes

$$\hat{K} = \frac{n\sum\limits_{i=1}^{k} n_{ii} - \sum\limits_{i=1}^{k} n_{i+}n_{+i}}{n^2 - \sum\limits_{i=1}^{k} n_{i+}n_{+i}} \; ; \quad n_{ii}, \; n_{i+}, \text{ and } n_{+i} \text{ as previously defined above.}$$

The approximate large sample variance of Kappa is computed using the Delta method as follows:

$$\text{vâr}\left(\hat{K}\right) = \frac{1}{n}\left\{ \frac{\theta_1(1-\theta_1)}{(1-\theta_2)^2} + \frac{2(1-\theta_1)(2\theta_1\theta_2 - \theta_3)}{(1-\theta_2)^3} + \frac{(1-\theta_1)^2(\theta_4 - 4\theta_2^2)}{(1-\theta_2)^4} \right\}$$

where

$$\theta_1 = \frac{1}{n}\sum_{i=1}^{k} n_{ii} \;,$$

$$\theta_2 = \frac{1}{n^2}\sum_{i=1}^{k} n_{i+}n_{+i} \;,$$

$$\theta_3 = \frac{1}{n^2}\sum_{i=1}^{k} n_{ii}\left(n_{i+} + n_{+i}\right),$$

and

$$\theta_4 = \frac{1}{n^3}\sum_{i=1}^{k}\sum_{j=1}^{k} n_{ij}\left(n_{j+} + n_{+i}\right)^2 \;.$$

A KHAT value is computed for each error matrix and is a measure of how well the remotely sensed classification agrees with the reference data. Confidence intervals around the KHAT value can be computed using the approximate large sample variance and the fact that the KHAT statistic is asymptotically normally distributed. This fact also provides a means for testing the significance of the KHAT statistic for a single error matrix to determine if the agreement between the remotely sensed classification and the reference data is significantly greater than 0 (i.e., better than a random classification).

It is always satisfying to see that your classification is meaningful and significantly better than a random classification. If it is not, you know that something has gone terribly wrong.

Finally, there is a test to determine if two independent KHAT values, and therefore two error matrices, are significantly different. With this test it is possible to statistically compare two analysts, two algorithms, or even two dates of imagery and see which produces the higher accuracy. Both of these tests of significance rely on the standard normal deviate as follows:

Let $\hat{K}_1$ and $\hat{K}_2$ denote the estimates of the Kappa statistic for error matrix #1 and #2, respectively. Let also $\text{vâr}\left(\hat{K}_1\right)$ and $\text{vâr}\left(\hat{K}_2\right)$ be the corresponding estimates of the variance as computed from the appropriate equations. The test statistic for testing the significance of a single error matrix is expressed by

$$Z = \frac{\hat{K}_1}{\sqrt{\text{vâr}\left(\hat{K}_1\right)}}.$$

Z is standardized and normally distributed (i.e., standard normal deviate). Given the null hypothesis $H_0{:}K_1 = 0$, and the alternative $H_1{:}K_1 \neq 0$, H_0 is rejected if $Z \geq Z_{\alpha/2}$, where $\alpha/2$ is the confidence level of the two-tailed Z test and the degrees of freedom are assumed to be ∞ (infinity).

The test statistic for testing if two independent error matrices are significantly different is expressed by

$$Z = \frac{\left|\hat{K}_1 - \hat{K}_2\right|}{\sqrt{\text{vâr}\left(\hat{K}_1\right) + \text{vâr}\left(\hat{K}_2\right)}}.$$

Z is standardized and normally distributed. Given the null hypothesis $H_0{:}(K_1 - K_2) = 0$, and the alternative $H_1{:}(K_1 - K_2) \neq 0$, H_0 is rejected if $Z \geq Z_{\alpha/2}$.

It is prudent at this point to provide an actual example so that the equations and theory can come alive to the reader. The error matrix presented as an example in Table 5-1 was generated from Landsat Thematic Mapper (TM) data using an unsupervised classification approach by analyst #1. A second error matrix was generated using the exact same imagery and same classification approach, however the clusters were labeled by analyst #2 (Table 5-3). It is important to note that analyst #2 was not as ambitious as analyst #1, and did not collect as much accuracy assessment data.

Table 5-4 presents the results of the Kappa analysis on the individual error matrices. The KHAT values are a measure of agreement or accuracy. The values can range from +1 to −1. However, since there should be a positive correlation between the remotely sensed classification and the reference data, positive KHAT values are expected. Landis and Koch (1977) characterized the possible ranges for KHAT into three groupings: a value greater than 0.80 (i.e., 80%) represents strong agreement; a value between 0.40 and 0.80 (i.e., 40–80%) represents moderate agreement; and a value below 0.40 (i.e., 40%) represents poor agreement.

Table 5-4 also presents the variance of the KHAT statistic and the Z statistic used for determining if the classification is significantly better than a random result. At the 95% confidence level, the critical value would be 1.96. Therefore, if the

Table 5-3   An Error Matrix Using the Same Imagery and Classification Algorithm as in Table 5-1 Except That the Work Was Done by a Different Analyst

		Reference Data				row total
		D	C	AG	SB	
	D	45	4	12	24	85
	C	6	91	5	8	110
Classified Data	AG	0	8	55	9	72
	SB	4	7	3	55	104
	column total	55	110	75	96	336

Land Cover Categories

D = deciduous

C = conifer

AG = agriculture

SB = shrub

OVERALL ACCURACY =
(45+91+55+55)/ 336 =
246/3 36 = 73%

PRODUCER'S ACCURACY

D = 45/5 5 = 82%
C = 91/1 10 = 83%
AG = 55/ 75 = 73%
SB = 55/9 6 = 57%

USER'S ACCURACY

D = 45/8 5 = 53%
C = 91/1 10 = 83%
AG = 55/ 72 = 76%
SB = 55/6 9 = 80%

Table 5-4   Individual Error Matrix Kappa Analysis Results

Error Matrix	KHAT	Variance	Z statistic
Analyst #1	0.65	0.0007778	23.4
Analyst #2	0.64	0.0010233	20.0

Table 5-5   Kappa Analysis Results for the Pairwise Comparison of the Error Matrices

Pairwise Comparison:	Z statistic
Analyst #1 vs. Analyst #2	0.3087

absolute value of the test Z statistic is greater than 1.96, the result is significant, and you would conclude that the classification is better than random. The Z statistic values for the two error matrices in Table 5-4 are both 20 or more, and so both classifications are significantly better than random.

Table 5-5 presents the results of the Kappa analysis that compares the error matrices, two at a time, to determine if they are significantly different. This test is based on the standard normal deviate and the fact that although remotely sensed data are discrete, the KHAT statistic is asymptotically normally distributed. The results of this pairwise test for significance between two error matrices reveals that these two matrices are not significantly different. This is not surprising since the overall accuracies were 74% and 73% and the KHAT values were 0.65 and 0.64, respectively. Therefore, it could be concluded that these two analysts may work together because they produce approximately equal classifications. If two different techniques or algorithms were being tested and if they were shown to be not significantly different, then it would be best to use the cheaper, quicker, or more efficient approach.

Margfit

In addition to the Kappa analysis, a second technique called Margfit can be applied to "normalize" or standardize the error matrices for comparison purposes. Margfit uses an iterative proportional fitting procedure which forces each row and column (i.e., marginal) in the matrix to sum to a predetermined value; hence the name Margfit. If the predetermined value is one, then each cell value is a proportion of one and can easily be multiplied by 100 to represent percentages. The predetermined value could also be set to 100 to obtain percentages directly or to any other value the analyst chooses.

In this normalization process, differences in sample sizes used to generate the matrices are eliminated and therefore, individual cell values within the matrix are directly comparable. In addition, because as part of the iterative process, the rows and columns are totaled (i.e., marginals), the resulting normalized matrix is more indicative of the off-diagonal cell values (i.e., the errors of omission and commission). In other words, all the values in the matrix are iteratively balanced by row and column, thereby incorporating information from that row and column into each individual cell value. This process then changes the cell values along the major diagonal of the matrix (correct classifications), and therefore a normalized overall accuracy can be computed for each matrix by summing the major diagonal and dividing by the total of the entire matrix.

Consequently, one could argue that the normalized accuracy is a better representation of accuracy than is the overall accuracy computed from the original matrix because it contains information about the off-diagonal cell values. Table 5-6 presents the normalized matrix generated from the original error matrix presented in Table 5-1 (an unsupervised classification of Landsat TM data by analyst #1) using the Margfit procedure. Table 5-7 presents the normalized matrix generated from the original error matrix presented in Table 5-3, which used the same imagery and classifier, but was performed by analyst #2.

In addition to computing a normalized accuracy, the normalized matrix can also be used to directly compare cell values between matrices. For example, we may be interested in comparing the accuracy each analyst obtained for the conifer category. From the original matrices we can see that analyst #1 classified 81 sample units correctly while analyst #2 classified 91 correctly. Neither of these numbers means

Table 5-6   Normalized Error Matrix from Analyst #1

Reference Data

		D	C	AG	SB
	D	0.7537	0.0261	0.1300	0.0909
Classified Data	C	0.1226	0.7735	0.0521	0.0517
	AG	0.0090	0.1042	0.7731	0.1133
	SB	0.1147	0.0962	0.0448	0.7440

3.0443

Land Cover Categories

D = deciduous
C = conifer
AG = agriculture
SB = shrub

NORMALIZED ACCURACY =
0.7537+0.7735+0.7731+0.7440 =
3.0443 / 4.0 = 76%

much, because they are not directly comparable due to the differences in the number of samples used to generate the error matrix by each analyst. Instead, these numbers would need to be converted into percentages or user's and producer's accuracies so that a comparison could be made.

Here another problem arises. Do we divide the total correct by the row total (user's accuracy) or by the column total (producer's accuracy)? We could calculate both and compare the results or we could use the cell value in the normalized matrix. Because of the iterative proportional fitting routine, each cell value in the matrix has been balanced by the other values in its corresponding row and column. This balancing has the effect of incorporating producer's and user's accuracies together. Also since each row and column add to one, an individual cell value can quickly be converted to a percentage by multiplying by 100. Therefore, the normalization process provides a convenient way of comparing individual cell values between error matrices regardless of the number of samples used to derive the matrix (Table 5-8).

Table 5-9 provides a comparison of the overall accuracy, the normalized accuracy, and the KHAT statistic for the two analysts. In this particular example, all three measures of accuracy agree about the relative ranking of the results. However, it is possible for these rankings to disagree simply because each measure incorpo-

Table 5-7   Normalized Error Matrix from Analyst #2

Reference Data

Classified Data		D	C	AG	SB
	D	0.7181	0.0312	0.1025	0.1488
	C	0.1230	0.7607	0.0541	0.0619
	AG	0.0136	0.1017	0.7848	0.0995
	SB	0.1453	0.1064	0.0587	0.6898

2.9534

Land Cover Categories

D = deciduous
C = conifer
AG = agriculture
SB = shrub

NORMALIZED ACCURACY =
0.7181+0.7607+0.7848+0.6898 =
2.9534 / 4.0 = 74%

rates various levels of information from the error matrix into its computations. Overall accuracy only incorporates the major diagonal and excludes the omission and commission errors. As already described, normalized accuracy directly includes the off-diagonal elements (omission and commission errors) because of the iterative proportional fitting procedure. As shown in the KHAT equation, KHAT accuracy indirectly incorporates the off-diagonal elements as a product of the row and column marginals. Therefore, depending on the amount of error included in the matrix, these three measures may not agree.

It is not possible to give clearcut rules as to when each measure should be used. Each accuracy measure incorporates different information about the error matrix and therefore must be examined as different computations attempting to explain the error. Our experience has shown that if the error matrix tends to have a great many off-diagonal cell values with zeros in them, then the normalized results tend to disagree with the overall and Kappa results.

Many zeros occur in a matrix when an insufficient sample has been taken or when the classification is exceptionally good. Because of the iterative proportional fitting routine, these zeros tend to take on positive values in the normalization process showing that some error could be expected. The normalization process then tends to reduce the accuracy because of these positive values in the off-diagonal cells. If

Table 5-8   Comparison of the Accuracy Values for an Individual Category

Error matrix	original cell value	producer's accuracy	user's accuracy	normalized value
Analyst #1	81	79%	81%	77%
Analyst #2	91	83%	83%	76%

Table 5-9   Summary of the Three Accuracy Measures for Analyst #1 and #2

Error matrix	Overall accuracy	KHAT	Normalized accuracy
Analyst #1	74 %	65%	76%
Analyst #2	73%	64%	74%

a large number of off-diagonal cells do not contain zeros then the results of the three measures tend to agree. There are also times when the Kappa measure will disagree with the other two measures. Because of the ease of computing all three measures and because each measure reflects different information contained within the error matrix, we recommend an analysis such as the one performed here to glean as much information from the error matrix as possible.

Conditional Kappa

In addition to computing the Kappa coefficient for an entire error matrix, it may be useful to look at the agreement for an individual category within the matrix. Individual category agreement can be tested using the conditional Kappa coefficient. The maximum likelihood estimate of the Kappa coefficient for conditional agreement for the ith category is given by

$$\hat{K}_i = \frac{nn_{ii} - n_{i+}n_{+i}}{nn_{i+} - n_{i+}n_{+i}}, \quad n_{i+} \text{ and } n_{+i} \text{ as previously defined above,}$$

and the approximate large sample variance for the ith category is estimated by

$$\hat{\text{var}}\left(\hat{K}_i\right) = \frac{n\left(n_{i+} - n_{ii}\right)}{\left[n_{i+}\left(n - n_{+i}\right)\right]^3}\left[\left(n_{i+} - n_{ii}\right)\left(n_{i+}n_{+i} - nn_{ii}\right) + nn_{ii}\left(n - n_{i+} - n_{+i} + n_{ii}\right)\right].$$

The same comparison tests available for the Kappa coefficient apply to this conditional Kappa for an individual category.

Weighted Kappa

The Kappa analysis is appropriate when all the error in the matrix can be considered of equal importance. However, it is easy to imagine a classification scheme where errors may vary in their importance. In fact, this latter situation is really the more realistic approach. For example, it may be far worse to classify a forested area as water than to classify it as shrub. In this case, the ability to weight the Kappa analysis would be very powerful (Cohen 1968). The following section describes the procedure to conduct a weighted Kappa analysis.

Let w_{ij} be the weight assigned to the i,jth cell in the matrix. This means that the proportion p_{ij} in the i,jth cell is to be weighted by w_{ij}. The weights should be restricted to the interval $0 \leq w_{ij} \leq 1$ for $i \neq j$ and the weights representing the maximum agreement are equal to 1, i.e., $w_{ii} = 1$ (Fleiss et al. 1969).

Therefore, let

$$p_o^* = \sum_{i=1}^{k} \sum_{j=1}^{k} w_{ij} p_{ij}$$

be the weighted agreement, and

$$p_c^* = \sum_{i=1}^{k} \sum_{j=1}^{k} w_{ij} p_{i+} p_{+j} \, , \quad p_{ij}, \ p_{i+}, \text{ and } p_{+j} \text{ as previously defined above,}$$

the weighted "chance agreement."

Then the weighted Kappa is defined by

$$\hat{K}_w = \frac{p_o^* - p_c^*}{1 - p_c^*} .$$

To compute the large sample variance of the weighted Kappa define the weighted average of the weights in the ith category of the remotely sensed classification by

$$\overline{w}_{i+} = \sum_{j=1}^{k} w_{ij} p_{+j} \, , \quad p_{+j} \text{ as previously defined above,}$$

and the weighted average of the weights in the jth category of the reference data set by

$$\overline{w}_{+j} = \sum_{i=1}^{k} w_{ij} p_{i+} \, , \quad p_{i+} \text{ as previously defined above.}$$

The variance may be estimated by

$$\text{v\^ar}\left(\hat{K}_w\right)=\frac{1}{n\left(1-p_c^*\right)^4}\left\{\sum_{i=1}^{k}\sum_{j=1}^{k}p_{ij}\left[w_{ij}\left(1-p_c^*\right)-\left(\overline{w}_{i+}+\overline{w}_{+j}\right)\left(1-p_o^*\right)\right]^2\right.$$

$$\left.-\left(p_o^*p_c^*-2p_c^*+p_o^*\right)^2\right\}.$$

The same tests of significant difference described previously for the Kappa analysis apply to the weighted Kappa. An individual weighted Kappa value can be evaluated to see if the classification is significantly better than random. Two independent weighted Kappas can also be tested to see if they are significantly different.

Although the weighted Kappa has been in the literature since the 1960s and even suggested to the remote sensing community by Rosenfield and Fitzpatrick-Lins (1986), it has not received widespread attention. The reason for this lack of use is undoubtedly the need to select appropriate weights. Manipulating the weighting scheme can significantly change the results. Therefore, comparisons between different projects using different weighting schemes would be very difficult. The subjectivity of choosing the weights is always hard to justify. Using the unweighted Kappa analysis avoids these problems.

Compensation for Chance Agreement

Some researchers and scientists have objected to the use of the Kappa coefficient for assessing the accuracy of remotely sensed classifications because the degree of chance agreement may be over-estimated (Foody 1992). Remember from the equation for computing the Kappa coefficient,

$$\hat{K}=\frac{p_o-p_c}{1-p_c},$$

that p_o is the observed proportion of agreement (i.e., the actual agreement) and p_c is the proportion of agreement that is expected to occur by chance (i.e., the chance agreement). However, in addition to the chance agreement, p_c also includes some actual agreement (Brennan and Prediger 1981) or agreement for cause (Aickin 1990). Therefore, since the chance agreement term does not consist solely of chance agreement, the Kappa coefficient may underestimate the classification agreement.

This problem is known to occur when the marginals are free (not fixed *a priori*), which is most often the case with remotely sensed classifications. Foody (1992) presents a number of possible solutions to this problem including two Kappa-like coefficients that compensate for chance agreement in different ways. However, given the very powerful properties of the Kappa coefficient, including the ability to test

Table 5-10   Error Matrix Showing Map Marginal Proportions

True (i)

Reference Data

		D	C	AG	SB	row total	map marginal proportions, π_j
	D	65	4	22	24	115	0.3
Map (j)	C	6	81	5	8	100	0.4
Classified Data	AG	0	11	85	19	115	0.1
	SB	4	7	3	90	104	0.2
column total		75	103	115	141	434	

OVERALL ACCURACY =
(65+81+85+90)/ 434 =
321/4 34 = 74%

for significant differences between two independent coefficients, it must still be considered a vital accuracy assessment measure.

Confidence Limits

Confidence intervals are extremely common and are an expected component of any statistical estimate. However, computing confidence intervals for values in an error matrix are more complex than simply computing a confidence interval for a traditional statistical analysis. The following example illustrates the calculations derived from the error matrix (Card 1982). This example is designed assuming simple random sampling. If another sampling scheme is used the variance equations change slightly.

The same error matrix as in Table 5-1 will be used to compute the confidence intervals. However, the map marginal proportions, π_j, computed as the proportion of the map falling into each map category, are also required (Table 5-10). The map marginal proportions are not derived from the error matrix, but are simply the proportion of the total map area falling into each category. These proportions can quickly be obtained by dividing the area of each category by the total map area.

Given this matrix, the first step is to compute the individual cell probabilities using the following equation:

$$\hat{p}_{ij} = \pi_j n_{ij} / n_{.j} .$$

The individual cell probabilities are simply the map marginal proportion multiplied by the individual cell value all divided by the row marginal. The results of these computations are shown in Table 5-11.

Table 5-11   Error Matrix of Individual Cell Probabilities, $\hat{p}_{ij}$

True (i)

Reference Data

		D	C	AG	SB
Map (j)	D	0.170	0.101	0.057	0.063
	C	0.024	0.324	0.020	0.032
Classified Data	AG	0.000	0.010	0.074	0.017
	SB	0.008	0.013	0.006	0.173

The true marginal proportions, $\hat{p}_i$, can then be computed using the equation

$$\hat{p}_j = \sum_{j=1}^{r} \pi_j n_{ij} / n_{\cdot j} \,.$$

The true marginal proportions can also be computed simply by summing the individual cell probabilities in each column. For example, $\hat{p}_1 = 0.170 + 0.024 + 0.000 + 0.008 = 0.202$, $\hat{p}_2 = 0.357$, $\hat{p}_3 = 0.157$, and $\hat{p}_4 = 0.285$.

The third step is to compute the probability correct given the true class i; in other words, the producer's accuracy. It should be noted that the values here differ somewhat from those computed in the error matrix discussion because these values have been corrected for bias by incorporating the true marginal proportions as shown in the following equation:

$$\hat{\theta}_{ii} = \left(\pi_i / \hat{p}_i\right)\left(n_{ii} / n_{\cdot i}\right) \quad \text{or} \quad \hat{p}_{ii} / \hat{p}_i \,.$$

As expected, the producer's accuracy is computed taking the diagonal cell value from the cell probability matrix (Table 5-11) and dividing by the true marginal proportion. For example, $\theta_{11} = 0.170/0.202 = 0.841$ or 84%, $\theta_{22} = 0.908$, $\theta_{33} = 0.471$, and $\theta_{44} = 0.607$.

The next step is to compute the probability correct given map class j; in other words, the user's accuracy. This computation is made exactly as described in the error matrix discussion by taking the diagonal cell value and dividing by the row (j) marginal. The equation for this calculation is as follows:

$$\hat{l}_{jj} = n_{jj} / n_{\cdot j} \,.$$

Therefore, $\hat{l}_{11} = 65/115 = 0.565$ or 57%, $\hat{l}_{22} = 0.810$, $\hat{l}_{33} = 0.739$, and $\hat{l}_{44} = 0.865$.

Step 5 is to compute the overall correct by summing the major diagonal of the cell probabilities or using the equation

$$\hat{P}_c = \sum_{j=1}^{r} \pi_j n_{jj}/n_{\cdot j} \,.$$

Therefore, in this example, $\hat{P}_c = 0.170 + 0.324 + 0.074 + 0.173 = 0.741$ or 74%.

We have now made essentially the same calculations as described in the error matrix discussion except that we have corrected for bias by using the true marginal proportions. The next step is to compute the variances for those terms (overall, producer's and user's accuracies) that we wish to calculate confidence intervals.

Variance for overall accuracy, $\hat{P}_c$,

$$V\!\left(\hat{P}_c\right) = \sum_{i=1}^{r} p_{ii}\!\left(\pi_i - p_{ii}\right)\!/\!\left(\pi_i n\right) .$$

Therefore, in this example,

$$\hat{P}_c = \big[0.170(0.3 - 0.170)/(0.3)(434)$$

$$+ 0.324(0.4 - 0.324)/(0.4)(434)$$

$$+ 0.074(0.1 - 0.074)/(0.1)(434)$$

$$+ 0.173(0.2 - 0.173)/(0.2)(434)$$

$$= 0.00040.$$

Confidence interval for overall accuracy, $\hat{P}_c$,

$$\hat{P}_c = 2\!\left[V\!\left(\hat{P}_c\right)\right]^{1/2} .$$

Therefore, in this example, the confidence interval for

$$\hat{P}_c = 0.741 \pm 2(0.0004)^{1/2}$$

$$= 0.741 \pm 2(0.02)$$

$$= 0.741 \pm 0.04$$

$$= (0.701,\ 0.781) \quad \text{or} \quad 70\% \text{ to } 78\%.$$

Variance for producer's accuracy, $\hat{\theta}_{ii}$,

$$V(\hat{\theta}_{ii}) = p_{ii}p_i^{-4}\left[p_{ii}\sum_{j\neq 1}^{r} p_{ij}(\pi_j - p_{ij})/\pi_j n + (\pi_j - p_{ii})(p_i - p_{ii})^2/\pi_i n\right].$$

Therefore, in this example,

$$V(\hat{\theta}_{11}) = 0.170(0.202)^{-4}\{0.170[0.024(0.4 - 0.024)$$

$$/(0.4)(434) + 0.008(0.2 - 0.008)/(0.2)(434)]$$

$$+ (0.3 - 0.170)(0.202 - 0.170)^2/(0.3)(434)\}$$

$$= 0.00132.$$

Confidence interval for producer's accuracy, $\hat{\theta}_{ii}$,

$$\hat{\theta}_{ii} \pm 2\left[V(\hat{\theta}_{ii})\right]^{1/2}.$$

Therefore, in this example, the confidence interval for

$$\hat{\theta}_{11} = 0.841 \pm 2(0.00132)^{1/2}$$

$$= 0.841 \pm 2(0.036)$$

$$= 0.841 \pm 0.072$$

$$= (0.768,\ 0.914) \quad \text{or} \quad 77\% \text{ to } 91\%.$$

Variance for user's accuracy, $\hat{l}_{ii}$,

$$V(\hat{l}_{ii}) = p_{ii}(\pi_i - p_{ii})/\pi_i^2 n.$$

Therefore, in this example,

$$V(\hat{l}_{11}) = 0.170(0.3 - 0.170)/(0.3)^2(434)$$

$$= 0.00057.$$

Confidence interval for

$$\hat{l}_{ii} \pm 2\left[V(\hat{l}_{ii})\right]^{1/2}$$

Therefore, in this example, the confidence interval for

$$\hat{l}_{11} = 0.565 \pm 2(0.00057)^{1/2}$$

$$= 0.565 \pm 2(0.024)$$

$$= 0.741 \pm 0.048$$

$$= (0.517, 0.613) \quad \text{or} \quad 52\% \text{ to } 61\%.$$

It must be remembered that these confidence intervals are computed from asymptotic variances. If the normality assumption is valid, then these are 95% confidence intervals. If not, then by Chebyshev's inequality, they are at least 75% confidence intervals.

Area Estimation/Correction

In addition to all the uses of an error matrix already presented, it can also be used to update the areal estimates of the map categories. The map derived from the remotely sensed data is a complete enumeration of the ground. However, the error matrix is an indicator of where misclassification occurred between what the map said and what is actually on the ground. Therefore, it is possible to use the information from the error matrix to revise the estimates of total area for each map category. It is not possible to update the map itself or to revise a specific location on the map, but it is possible to revise total area estimates. Updating in this way may be especially important for small, rare categories whose estimates of total area could vary greatly depending on even small misclassification errors.

Czaplewski and Catts (1990) and Czaplewski (1992) have reviewed the use of the error matrix to update the areal estimates of map categories. They propose an informal method, both numerically and graphically, to determine the magnitude of bias introduced in the areal estimates by the misclassification. They also review two methods of statistically calibrating the misclassification bias. The first method is called the classifical estimator and was proposed to the statistical community by Grassia and Sundberg (1982) and used in a remotely sensed application by Prisley and Smith (1987) and Hay (1988). The classical estimator uses the probabilities from the omission errors for calibration.

The second method is the inverse estimator, which uses the probabilities from the commission errors to calibrate the areal estimates. Tenenbein (1972) introduced this technique in the statistical literature and Chrisman (1982) and Card (1982) have used it for remote sensing applications. The confidence calculations derived in the previous section are from Card's (1982) work using the inverse estimator for calibration. More recently, Woodcock (1996) has proposed a modification of the Card approach incorporating fuzzy set theory into the calibration process.

Despite all this work, not many users have picked up on these calibration techniques or the need to perform the calibration. From a practical standpoint, overall total areas are not that important. We have already discussed this in terms of non-site specific accuracy assessment. However, as more and more work is done with looking at change, and especially changes of small, rare categories, the use of these calibration techniques may gain in importance.

CHAPTER 6

Analysis of Differences in the Error Matrix

After testing the error matrix for statistical significance, the next step in analysis involves discovering why some of the accuracy map site labels do not match the reference labels. While much attention is placed on overall accuracy percentages, by far the more interesting analysis concerns learning why sites do not fall on the diagonal of the error matrix. To both effectively use the map and to make better maps in the future, we need to know what causes the differences in the matrix.

All differences will be the result of one of four possible sources:

1. Errors in the reference data;
2. Sensitivity of the classification scheme to observer variability;
3. Inappropriateness of the remote sensing technology for mapping a specific land cover class; and
4. Mapping error.

This chapter reviews each one of these sources and discusses the impacts of each one to accuracy assessment results.

ERRORS IN THE REFERENCE DATA

A major assumption of the error matrix is that the label from the reference information represents the "true" label of the site and that all differences between the remotely sensed map classification and the reference data are due to classification and/or delineation error. Unfortunately, error matrices can be inadequate indicators of map error, because they are often confused by errors in the reference data (Congalton and Green, 1993), a function of

- Registration differences between the reference data and the remotely sensed map classification caused by delineation and/or digitizing errors. For example, if GPS is not used in the field during accuracy assessment, it is possible for field personnel to collect data in the wrong area. Other registration errors can occur when an accuracy assessment site is incorrectly delineated or digitized, or when an existing map used for reference data is not precisely registered to the map being assessed.

- Data entry errors. Data entry errors are common in any database project and can be controlled only through rigorous quality control. Developing digital data entry forms that will only allow a certain set of characters for specific fields can catch errors during data entry. One of the best (yet most expensive) methods for catching data entry errors is to enter all data twice and then compare the two data sets. Differences usually indicate an error.

- Classification scheme errors. Every accuracy assessment map and reference site must have a label derived from the classification scheme used to create the map. Classification scheme errors occur when personnel misapply the classification scheme to the map or reference data; a common occurrence with complex classification schemes. If the reference data is in a database, then such errors can be avoided or at least highlighted, by programming the classification scheme rules, and using the program to label accuracy assessment sites. Classification scheme errors also occur when the classification scheme used to label the reference site is different from the one used to create the map—a common occurrence when existing data or maps are used as reference data.

- Changes in land cover between the date of the remotely sensed data and the date of the reference data. As the second section of Chapter 4 details, landcover change can have a profound effect on accuracy assessment results. Tidal differences, crop or tree harvesting, urban development, fire, and pests all can cause the landscape to change in the time period between capturing the remotely sensed data and accuracy assessment reference data collection.

- Mistakes in labeling reference data. Labeling mistakes usually occur because inexperienced personnel are used to collect reference data. Even with experienced personnel, the more detailed the classification scheme the more likely an error will occur. Some conifer and hardwood species are difficult to distinguish on the ground, much less from aerial photography. Young crops of broccoli, Brussels sprouts, and cauliflower are easily confused. Thus, accuracy assessment must also be completed on the reference data. If photo interpretation is used to assess a map from satellite imagery, then a sample of the photo-interpreted sites must be visited on the ground. If only field data is used, then some of the sites must be visited twice by two different personnel.

Table 6.1 summarizes reference data errors discovered during quality control of a recent assessment. Only six of the differences between the map and reference labels were caused by errors in the map. Over two thirds of the differences (85 sites) were caused by mistakes in the reference data. The most significant error occurred from using different classification schemes (50 sites). In this project, National Wetlands Inventory (NWI) maps were used exclusively to map wetlands, i.e., wetlands were defined in the classification scheme to be those areas identified by NWI data as wetlands. However, when accuracy assessment was done, the reference photo interpreters used a different definition of wetlands. The remaining differences were caused by observer variation, discussed in the next section of this chapter.

Table 6-1 Analysis of Map and Reference Label Differences

Map vs. Ref. Difference	Number of Sites Different	Map Error	Reference Label Error	Date Change	Class. Scheme Difference	Variation in Estimation
Barren vs. Water	19	0	6	8	0	5
Hardwood vs. Water	6	0	0	0	0	6
Herb vs. Forested	50	6	17	4	0	23
Wetland vs. All Other Types	50	0	0	0	50	0
Total	125	6	23	12	50	34

SENSITIVITY OF THE CLASSIFICATION SCHEME TO OBSERVER VARIABILITY

Classification scheme rules often impose discrete boundaries on continuous conditions in nature such as vegetation cover. In situations where classification scheme breaks represent artificial distinctions along a continuum, observer variability is often difficult to control and, while unavoidable, can have profound effects on accuracy assessment results (Congalton, 1991; Congalton and Green, 1993). Analysis of the error matrix must include investigations concerning how much of the matrix difference results from observers being unable to precisely distinguish between classes when the accuracy assessment site is on the margin between two or more classes in the classification scheme.

Plato's allegory in the cave is useful for thinking about observer variability. In the allegory, Plato describes prisoners who cannot move:

> Above and behind them a fire is blazing in the distance, and between the fire and the prisoners there is a ... screen which marionette players have in front of them over which they show puppets ... [The prisoners] see only their own shadows, or the shadows of one another which the fire throws on the opposite wall of the cave .... To them ... the truth would be literally nothing but the shadows of the images. (Plato, *The Republic*, Book VII, 515-B, from Benjamin Jowett's translation as published in Vintage Classics, Random House, New York, 1991.)

Like Plato's prisoners in the cave, we all perceive the world within the context of our experience. The difference between reality and perceptions of reality is often as fuzzy as Plato's shadows. Between ourselves and from day to day, our observations and perceptions vary depending on our training, experience, or mood.

The analysis in Table 6.1 shows the impact that variation in interpretation can have on accuracy assessment. In the project, two photo interpreters were asked to label the same accuracy assessment reference sites. Almost 30% of the differences between the map and reference label were caused by variation in interpretation.

Consider, for example, the assessment of a map of tree crown closure with classification scheme rules defining classes as

Unvegetated   0–10%,
Sparse   11–30%,
Light   31–50%,
Medium   51–70%
Heavy   71–100%.

An accuracy assessment reference site from photo interpretation estimated at 45% tree crown cover could feasibly be considered correct with either a Light or Medium label because photo interpretation can be ±10%. The map user would be more concerned with a difference caused by a map label of Unvegetated versus a reference label of Heavy tree crown cover. Differences on class margins are both inevitable and far less significant to the map user than other types of differences.

Classification systems sensitive to estimates of vegetative cover are particularly susceptible to this type of confusion in the error matrix. Appendix 1 of this chapter shows the very complex classification scheme rules for a recently completed mapping project of Wrangell–St. Elias National Park in Alaska. The classification scheme is extremely affected by estimates of percent vegetative cover. Sensitivity analysis on 140 accuracy assessment sites revealed that nearly 33% of the sites received new class labels when estimates of vegetative cover were varied by as little as 5%.

Several researchers have noted the impact of the variation in human interpretation on map results and accuracy assessment (Gong and Chen, 1992; Lowell, 1992; Congalton and Biging, 1992; Congalton and Green, 1993). Gopal and Woodcock (1994) state, "The problem that makes accuracy assessment difficult is that there is ambiguity regarding the appropriate map label for some locations. The situation of one category being exactly right and all other categories being equally and exactly wrong often does not exist." Lowell (1992) calls for "a new model of space which shows transition zones for boundaries, and polygon attributes as indefinite." As Congalton and Biging (1992) conclude in their study of the validation of photo interpreted stand type maps, "The differences in how interpreters delineated stand boundaries was most surprising. We were expecting some shifts in position, but nothing to the extent that we witnessed. This result again demonstrates just how variable forests are and the subjectiveness of photo interpretation."

While it is difficult to control observer variation, it is possible to measure the variation and to use the measurements to compensate for differences between reference and map data that are caused not by map error but by variation in interpretation. One option is to measure each reference site precisely to reduce observer variance in reference site labels. This method can be prohibitively expensive, usually requiring extensive field sampling. The second option incorporates fuzzy logic into the reference data to compensate for non-error differences between reference and map data and is discussed in Chapter 7.

INAPPROPRIATENESS OF THE REMOTE SENSING TECHNOLOGY

Early satellite remote sensing projects were primarily concerned with testing the viability of various remote sensing technologies for mapping certain types of land cover. Researchers tested the hypotheses of whether or not a technology could be used to detect land use, crop types, or forest types. Many accuracy assessment techniques were developed primarily to test these hypotheses.

Recent accuracy assessment is more focused on learning about the reliability of a map for land management or policy analysis. However, some of the differences in the error matrix will be because the map producer was attempting to use a remote sensing technology that was incapable of distinguishing certain class types. Understanding what differences are caused by the technology is useful to the map producer when the next map is being made.

In the Wrangell–St. Elias example cited above, Landsat TM data was employed as the primary remotely sensed data, with 1:60,000 aerial photography as ancillary data. The classification scheme included distinctions between pure and mixed stands of black and white spruce. Accuracy assessment analysis showed consistent success at differentiating *pure* stands of black versus white spruce. However, consistently differentiating these species in mixed or occasional hybrid stands was found to be unreliable. This phenomenon is not surprising considering the difficulty often associated with differentiating these species in mixed and hybrid stands from the ground. In other words, remotely sensed data cannot be used to reliably differentiate these two types of conditions.

To make the map more reliable, the map user can collapse the classification system across classes. In this example, the non-pure spruce classes of Closed, Open, and Woodland were collapsed into an Unspecified Interior Spruce class. In the difference matrix, Unspecified Interior Spruce map labels were considered to be mapped correctly if they corresponded to a pure or mixed white spruce or black spruce reference site demonstrating the same density class of Closed, Open, or Woodland. For example, a map label of Open Unspecified Interior Spruce was considered to be correctly mapped if its corresponding reference label for the site was Open Black Spruce, Open White Spruce, or Open Black/White Spruce mix. While less information is displayed on the map, the remaining information is more reliable.

MAPPING ERROR

The final cause of differences in error matrices are the result of mapping error. Often these are difficult to distinguish from an inappropriate use of remote sensing technology. Usually, they are errors that are particularly obvious and unacceptable. For example, it is not uncommon for an inexperienced remote sensing professional to produce a map of land cover from satellite data that misclassifies northeast facing forests on steep slopes as water. Because water and shadowed wooded slopes both absorb most energy, this type of error is explainable, but unacceptable and avoidable. Many map users will be appalled at this type of error and are not particularly interested in having the electromagnetic spectrum explained to them. However, careful editing and comparison to aerial photography, checking that all water exists in areas without slope, and comparison to existing maps of waterways and lakes will all reduce the possibility of this type of map error.

Understanding the causes of this type of error can point the map producer to additional methods to improve the accuracy of the map. Perhaps other bands or band combinations will improve accuracy. Incorporation of ancillary data such as slope or elevation may be useful. In the Wrangell–St. Elias example, confusion existed between the Dwarf Shrub classes and the Graminoid class. The confusion was addressed through the use of unsupervised classifications and park-wide models utilizing digital elevation data, field-based data, and aerial photography. First, an unsupervised classification with 20 classes was executed for only those areas of the imagery classified as Dwarf Shrub in the map. A digital elevation coverage was utilized to stratify the study area for subsequent relabeling of unsupervised classes previously mapped as Dwarf Shrub but actually representing areas of Graminoid cover on the ground. From the unsupervised classification, two spectral classes were found to consistently represent Graminoid cover throughout the study area, while another spectral class was found to represent Graminoid cover in areas below 3,500 feet elevation. These spectral classes were subsequently recoded to the Graminoid class.

SUMMARY

Analysis of the causes of differences in the error matrix can be one of the most important steps in the creation of a map from remotely sensed data. In the past, too much emphasis has been placed on the overall accuracy of the map, without delving into the conditions that give rise to that accuracy. By understanding what causes the reference and map data to differ, we can use the map more reliably, and produce both better maps and better accuracy assessments in the future.

Appendix 1

WRANGELL–ST. ELIAS NATIONAL PARK AND PRESERVE LAND COVER MAPPING CLASSIFICATION KEY

If tree total ≥ 10% (Forested)
    If Conifer ≥ 75% of tree total
        If (Pigl + Pima) ≥ 67% of conifer total
            If (Pigl/(Pigl+Pima)) ≥ 75%   **PIGL**
            If (Pima/(Pigl+Pima)) ≥ 75%   **PIMA**
            Else   **Unspecified Spruce**
        If Broadleaf ≥ 75% of tree total   **Broadleaf**
        Else (mixed conifer/broadleaf)   **Spruce/Broadleaf**

Else If shrub total ≥ 25% (Shrub)
    If tall shrub total ≥ 25%   **Tall Shrub**
    If low shrub total ≥ 25%   **Low Shrub**
    If dwarf shrub total ≥ 25%   **Dwarf Shrub**
    Else (tall, low, or dwarf are not individually > 25%)
    If tall shrub total ≥ 67% of shrub total   **Tall Shrub**
    If low shrub total ≥ 67% or shrub total   **Low Shrub**
    If dwarf shrub total ≥ 67% of shrub total   **Dwarf Shrub**
    Else "pick the largest percent of":
        tall shrub   **Tall Shrub**
        low shrub   **Low Shrub**
        dwarf shrub   **Dwarf Shrub**
        (ties go to the "tallest")

Else if herbaceous ≥ 15% (Herbaceous)
    If graminoid ≥ 50% or (graminoid/herb total) ≥ 50%   **Graminoid**
    Else if forb ≥ 50% or (forb/herb total) ≥ 50%   **Forb**
    Else if moss ≥ 50% or (moss/herb total) ≥ 50%   **Moss/Lichen**
    Else if lichen ≥ 50% or (lichen/herb total) ≥ 50%   **Moss/Lichen**
    Else "pick the largest percent of":
        graminoid
        forb
        moss
        lichen
        (preference for ties go in the order listed)

Else if total vegetation ≥ 10% and < 30%   **Sparse Vegetation**

Else (non-vegetated)

    Water
    Barren
    Glacier/Snow
    Clouds/Cloud
    Shadow

WRANGELL–ST. ELIAS NATIONAL PARK AND PRESERVE
LAND COVER MAPPING CLASSES

Forested (>10% tree cover)
  Conifer (>75% conifer)
    Closed (60–100%)
      Pigl
      Pima
      Pigl/Pima
      Pisi
      Tshe
      Tsme
      Pisi/Tsme
      Pisi/Tshe
      Tshe/Tsme
      Spruce
      Mixed conifer
    Open (25–59%)
      Pigl
      Pima
      Pigl/Pima
      Pisi
      Tshe
      Tsme
      Pisi/Tsme
      Pisi/Tshe
      Tshe/Tsme
      Spruce
      Mixed conifer
    Woodland (10–24%)
      Pigl
      Pima
      Pigl/Pima
      Pisi
      Tshe
      Tsme
      Pisi/Tsme
      Pisi/Tshe
      Tshe/Tsme
      Spruce
      Mixed conifer
  Broadleaf (>75% broadleaf)
    Closed (60–100%)
      Closed Broadleaf
    Open (10–59%)
      Open Broadleaf
  Mixed
    Closed (60–100%)
      Pigl/Pima-Broadleaf
      Pisi-Broadleaf

    Tshe-Broadleaf
    Conifer-Broadleaf
  Open (10–59%)
    Pigl/Pima-Broadleaf
    Pisi-Broadleaf
    Tshe-Broadleaf
    Conifer-Broadleaf

Shrub (>25% shrub)
  Tall (tall shrub > 25% or dominant)
    Closed (>75%)
    Open (25–74%)
  Low (low shrub > 25% or dominant)
    Closed (>75%)
    Open (25–74%)
  Dwarf (dwarf shrub > 25% or dominant)

Herbaceous (herbaceous > 15%)
    Graminoid
    Forb
    Moss
    Lichen

Sparse vegetation
    Sparse vegetation

Non-vegetated
    Water
    Barren
    Glacier/Snow
    Clouds/Cloud Shadow

CHAPTER 7

Advanced Topics

BEYOND THE ERROR MATRIX

As remote sensing projects have grown in complexity, so have the associated classification schemes. The classification scheme then becomes a very important factor influencing the accuracy of the entire project. Recently, papers have appeared in the literature that point out some of the limitations of using only an error matrix approach to accuracy assessment with a complex classification scheme. A paper by Congalton and Green (1993) recommends the error matrix as a jumping off point for identifying sources of confusion (i.e., differences between the remotely sensed map and the reference data) and not just error in the remotely sensed classification. For example, the variation in human interpretation can have a significant impact on what is considered correct and what is not. As previously mentioned, if photo interpretation is used as the reference data in an accuracy assessment and that photo interpretation is not completely correct, then the results of the accuracy assessment will be very misleading. The same statements are true if ground observations, as opposed to actual ground measurements, are made and used as the reference data set. As classification schemes become more complex, more variation in human interpretation is introduced. Also, factors beyond just variation in interpretation are important. Work is needed to go beyond the error matrix and introduce techniques that build upon the information in the matrix and make it more meaningful.

Some of this work has already begun. In situations where the breaks (i.e., divisions between classes) in the classification system represent artificial distinctions along a continuum, variation in human interpretation is often very difficult to control and, while unavoidable, can have profound effects on accuracy assessment results (Congalton 1991, Congalton and Green 1993). Several researchers have noted the impact of the variation in human interpretation on map results and accuracy assessment (Gong and Chen 1992, Lowell 1992, McGuire 1992, Congalton and Biging 1992).

Gopal and Woodcock (1994) proposed the use of fuzzy sets to "allow for explicit recognition of the possibility that ambiguity might exist regarding the appropriate map label for some locations on the map. The situation of one category being exactly right and all other categories being equally and exactly wrong often does not exist."

In such an approach, it is recognized that instead of a simple system of correct (agreement) and incorrect (disagreement), there can be a variety of responses such as absolutely right, good answer, acceptable, understandable but wrong, and absolutely wrong.

Lowell (1992) calls for "a new model of space which shows transition zones for boundaries, and polygon attributes as indefinite." As Congalton and Biging (1992) conclude in their study of the validation of photo-interpreted stand-type maps, "the differences in how interpreters delineated stand boundaries was most surprising. We were expecting some shifts in position, but nothing to the extent that we witnessed. This result again demonstrates just how variable forests are and the subjectiveness of photo interpretation."

There are a number of methods that try to go beyond the basic error matrix in order to incorporate difficulties associated with building the matrix. These techniques all attempt to allow fuzziness into the assessment process and include modifying the error matrix, using fuzzy set theory, or measuring the variability of the classes.

Modifying the Error Matrix

The simplest method for allowing some consideration of the idea that class boundaries may be fuzzy is to accept as correct plus or minus one class of the actual class. This method works well if the classification is continuous such as tree size class or forest crown closure. If the classification is discrete vegetation classes, then this method may be totally inappropriate. Table 7-1 presents the traditional error matrix for a classification of forest crown closure. Only exact matches are considered correct and are tallied along the major diagonal. The overall accuracy of this classification is 40%. Table 7-2 presents the same error matrix, only the major diagonal has been expanded to include plus or minus one crown closure class. In other words, for crown closure class 3 both crown closure classes 2 and 4 are also accepted as correct. This revised major diagonal then results in a tremendous increase in overall accuracy to 75%.

The advantage of using this method of accounting for fuzzy class boundaries is obvious: the accuracy of the classification can increase dramatically. The disadvantage is that if the reason for accepting plus or minus one class cannot be adequately justified, then it may be viewed that you are cheating to try to get higher accuracies. Therefore, although this method is very simple to apply, it should be used only when everyone agrees it is a reasonable course of action. The other techniques described next may be more difficult to apply, but easier to justify.

Fuzzy Set Theory

Fuzzy set theory or fuzzy logic is a form of set theory. While initially introduced in the 1920s, fuzzy logic gained its name and its algebra in the 1960s and 1970s from Zadeh (1965), who developed fuzzy set theory as a way to characterize the ability of the human brain to deal with vague relationships. The key concept is that membership in a class is a matter of degree. Fuzzy logic recognizes that, on the margins of classes that divide a continuum, an item may belong to both classes. As Gopal and Woodcock (1994) state, "The assumption underlying fuzzy set theory is

Table 7-1   Error Matrix Showing the Ground Reference Data versus
the Image Classification for Forest Crown Closure

	ground reference						row total
	1	2	3	4	5	6	
1	2	9	1	2	1	1	16
2	2	8	3	6	1	1	21
3	0	3	3	4	9	1	20
4	0	0	2	8	7	10	27
5	0	1	2	1	6	16	26
6	0	0	0	0	3	31	34
column total	4	21	11	21	27	60	144

(left margin label: image classification)

Crown Closure
Categories

Class 1 =   0%   CC
Class 2 = 1 - 10% CC
Class 3 = 11 - 30% CC
Class 4 = 31 - 50% CC
Class 5 = 51 - 70% CC
Class 6 = 71 - 100% CC

OVERALL ACCURACY =
58/144 =   40%

PRODUCERS ACCURACY	USERS ACCURACY
Class 1 = 2/4 = 50%	Class 1 = 2/16 = 13%
Class 2 = 8/21 = 38%	Class 2 = 8/21 = 38%
Class 3 = 3/11 = 27%	Class 3 = 3/20 = 15%
Class 4 = 8/21 = 38%	Class 4 = 8/27 = 30%
Class 5 = 6/27 = 22%	Class 5 = 6/26 = 23%
Class 6 = 31/60 = 52%	Class 6 = 31/34 = 91%

that the transition from membership to non-membership is seldom a step function."
Therefore, while a 100% hardwood stand can be labeled hardwood, and a 100%
conifer stand may be labeled conifer, a 49% hardwood and 51% conifer stand may
be acceptable if labeled either conifer or hardwood.

A difficult task in using fuzzy logic is the development of rules for its application.
Fuzzy systems often rely on experts for the development of rules. Gopal and Woodcock (1994) relied on experts in their application of fuzzy sets to accuracy assessment
for Region 5 of the U.S. Forest Service. Their technique has been also successfully
applied by Pacific Meridian Resources in the assessment of forest type maps on the
Quinalt Indian Reservation as well as in the assessment of forest type maps for a
portion of the Tongass National Forest. Hill (1993) developed an arbitrary but
practical fuzzy set rule that determined "sliding class widths" for the assessment of
accuracy of maps produced for the California Department of Forestry and Fire
Protection of the Klamath Province in northwestern California.

Table 7-3 presents the results of a set of fuzzy rules applied to building the same
error matrix as was presented in Table 7-1. In this case, the rules were defined as
follows:

• Class 1 was defined as always 0% crown closure. If the reference data indicated
a value of 0%, then only an image classification of 0% was accepted.

Table 7-2 Error Matrix Showing the Ground Reference Data versus the Image Classification for Forest Crown Closure within Plus or Minus One Tolerance Class

	ground reference						row total
	1	2	3	4	5	6	
image classification 1	2	9	1	2	1	1	16
2	2	8	3	6	1	1	21
3	0	3	3	4	9	1	20
4	0	0	2	8	7	10	27
5	0	1	2	1	6	16	26
6	0	0	0	0	3	31	34
column total	4	20	11	21	27	60	144

Crown Closure Categories

Class 1 = 0% CC
Class 2 = 1 - 10% CC
Class 3 = 11 - 30% CC
Class 4 = 31 - 50% CC
Class 5 = 51 - 70% CC
Class 6 = 71 - 100% CC

OVERALL ACCURACY = 108/144 = 75%

PRODUCERS ACCURACY

Class 1 = 4/4 = 100%
Class 2 = 20/21 = 95%
Class 3 = 8/11 = 73%
Class 4 = 13/21 = 62%
Class 5 = 16/27 = 59%
Class 6 = 47/60 = 78%

USERS ACCURACY

Class 1 = 11/16 = 69%
Class 2 = 13/21 = 62%
Class 3 = 10/20 = 50%
Class 4 = 17/27 = 63%
Class 5 = 23/26 = 88%
Class 6 = 34/34 = 100%

- Class 2 was defined as acceptable if the reference data was within 5% of that of the image classification. In other words, if the reference data indicates that a sample has 15% crown closure and the image classification put it in Class 2, the answer would not be absolutely correct, but acceptable.
- Classes 3 through 6 were defined as acceptable if the reference data were within 10% of that of the image classification. In other words, a sample classified as Class 4 on the image but found to be 55% crown closure on the reference data would be considered acceptable.

As a result of these rules, off-diagonal elements in the matrix contain two separate values. The first value represents those that, although not absolutely correct, are acceptable within the fuzzy rules. The second value indicates those that are still unacceptable. Therefore, in order to compute the accuracies (overall, producer's, and user's), the values along the major diagonal and those deemed acceptable (i.e., those in the first value) in the off-diagonal elements are combined. In Table 7-3, this combination of absolutely correct and acceptable answers results in an overall accuracy of 64%. This overall accuracy is significantly higher than the original error matrix (Table 7-1), but not as high as that of Table 7-2.

It is much easier to justify the fuzzy rules used in generating Table 7-3 than it is to simply extend the major diagonal to plus or minus one whole class, as was done in Table 7-2. For crown closure it is recognized that mapping typically varies by plus or minus 10% (Spurr 1948). Therefore, it is reasonable to define as acceptable

Table 7-3 Error Matrix Showing the Ground Reference Data versus the Image Classification for Forest Crown Closure Using the Fuzzy Logic Rules

		ground reference					row total
		1	**2**	**3**	**4**	**5**	**6**	

image classification		1	2	3	4	5	6	row total
	1	2	6,3	1	2	1	1	16
	2	0,2	8	2,1	6	1	1	21
	3	0	2,1	3	4,0	9	1	20
	4	0	0	0,2	8	5,2	10	27
	5	0	1	2	1,0	6	12,4	26
	6	0	0	0	0	2,1	31	34
column total		4	21	11	21	27	60	144

Crown Closure Categories

Class 1 = 0% CC
Class 2 = 1 - 10% CC
Class 3 = 11 - 30% CC
Class 4 = 31 - 50% CC
Class 5 = 51 - 70% CC
Class 6 = 71 - 100% CC

OVERALL ACCURACY = 92/144 = 64%

PRODUCERS ACCURACY	USERS ACCURACY
Class 1 = 2/4 = 50%	Class 1 = 8/16 = 50%
Class 2 = 16/21 = 76%	Class 2 = 10/21 = 48%
Class 3 = 5/11 = 45%	Class 3 = 9/20 = 45%
Class 4 = 13/21 = 62%	Class 4 = 13/27 = 48%
Class 5 = 13/27 = 48%	Class 5 = 19/26 = 73%
Class 6 = 43/60 = 72%	Class 6 = 33/34 = 97%

a range within 10% for classes 3–6. Class 1 and Class 2 take an even more conservative approach and are therefore even easier to justify.

In addition to this fuzzy set theory working for continuous variables such as crown closure, it also applies to more categorical data. For example, in the hardwood range area of California many land cover types differ only by which hardwood species is dominant. In many cases, the same species are present and the specific land cover type is determined by which species is most abundant. Also, in some of these situations, the species look very much alike on aerial photography and on the ground. Therefore, the use of these fuzzy rules, which allow for acceptable answers as well as absolutely correct answers, makes a great deal of sense. It is easy to envision other examples that make use of this very powerful concept of absolutely correct and acceptable answers.

Measuring Variability

While it is difficult to control variation in human interpretation, it is possible to measure the variation and to use the measurements to compensate for differences between reference and map data that are caused not by map error but by variation in interpretation. There are two options available to control the variation in human interpretation to reduce the impact of this variation on map accuracy. One is to measure each reference site precisely to reduce variance in reference site labels. This method can be prohibitively expensive, usually requiring extensive field sampling. The second option measures the variance and uses the measurements to compensate for non-error differences between reference and map data. While the photo interpreter is an integral part of the process, an objective and repeatable method to capture the impacts of human variation is required. This technique is also time-consuming and expensive, as multiple interpreters must evaluate each accuracy assessment site. Presently, little work is being done to effectively evaluate variation in human interpretation.

COMPLEX DATA SETS

Change Detection

In addition to the difficulties associated with a single-date accuracy assessment of remotely sensed data, change detection presents even more difficult and challenging problems. For example, how does one obtain information on the reference data for images that were taken in the past, or how can one sample enough areas that will change in the future to have a statistically valid assessment, and which change detection technique will produce the best accuracy for a given change in the environment? Figure 7-1 is a modification of the sources of error figure presented at the beginning of this book (Figure 1-1) and shows how complicated the error sources get when performing a change detection. Most of the studies on change detection conducted up to this point do not present quantitative results of their work, which makes it difficult to determine which method should be applied to a future project.

All change detection techniques, except postclassification and direct multidate classification, use a threshold value to determine which pixels have changed from those pixels that have not changed. The threshold value can be determined as a standard deviation from the mean or chosen interactively (Fung and LeDrew 1988). Depending on the threshold value, very different accuracies can be obtained using the same change detection techniques. Fung and LeDrew (1988) developed a technique to determine the optimal threshold level. Using different threshold levels, they compared different classification accuracies in order to obtain the highest classification accuracy. Because all of the cells of the matrix are considered, the Kappa coefficient of agreement was the recommended measure of accuracy.

To date, no standard accuracy assessment technique for change detection has been developed. Studies on determining the optimal threshold value (Fung and LeDrew 1988) and the accuracies between different change detection techniques (Martin 1989, Singh 1986) have made encouraging steps toward accomplishing standard accuracy assessment techniques for change detection. However, as change

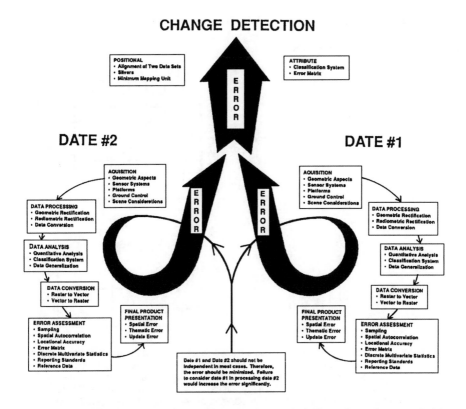

CHANGE DETECTION

DATE #2

DATE #1

POSITIONAL
• Alignment of Two Data Sets
• Slivers
• Minimum Mapping Unit

ATTRIBUTE
• Classification System
• Error Matrix

ERROR

AQUISITION
• Geometric Aspects
• Sensor Systems
• Platforms
• Ground Control
• Scene Considerations

DATA PROCESSING
• Geometric Rectification
• Radiometric Rectification
• Data Conversion

DATA ANALYSIS
• Quantitative Analysis
• Classification System
• Data Generalization

DATA CONVERSION
• Raster to Vector
• Vector to Raster

ERROR ASSESSMENT
• Sampling
• Spatial Autocorrelation
• Locational Accuracy
• Error Matrix
• Discrete Multivariate Statistics
• Reporting Standards
• Reference Data

FINAL PRODUCT
PRESENTATION
• Spatial Error
• Thematic Error
• Update Error

AQUISITION
• Geometric Aspects
• Sensor Systems
• Platforms
• Ground Control
• Scene Considerations

DATA PROCESSING
• Geometric Rectification
• Radiometric Rectification
• Data Conversion

DATA ANALYSIS
• Quantitative Analysis
• Classification System
• Data Generalization

DATA CONVERSION
• Raster to Vector
• Vector to Raster

ERROR ASSESSMENT
• Sampling
• Spatial Autocorrelation
• Locational Accuracy
• Error Matrix
• Discrete Multivariate Statistics
• Reporting Standards
• Reference Data

FINAL PRODUCT
PRESENTATION
• Spatial Error
• Thematic Error
• Update Error

Date #1 and Date #2 should not be
independent in most cases. Therefore,
the error should be minimized. Failure
to consider date #1 in processing date #2
would increase the error significantly.

Figure 7-1 Sources of error in a change detection analysis from remotely sensed data. Reproduced with permission, the American Society for Photogrammetry and Remote Sensing, from: Congalton, R.G. 1996. Accuracy assessment: A critical component of land cover mapping. IN: *Gap Analysis: A Landscape Approach to Biodiversity Planning.* A Peer-Reviewed Proceedings of the ASPRS/GAP Symposium. Charlotte, NC. pp. 119-131.

detection studies become more popular, the urgency for procedures to determine the accuracy the different techniques becomes increasingly important.

In order to apply the established accuracy assessment techniques to change detection, the standard classification error matrix needs to be adapted to a change detection error matrix. This new matrix has the same characteristics of the classification error matrix, but also assesses errors in changes between two time periods and not simply a single classification. An example (Figure 7-2) demonstrates the use of a change detection error matrix.

Figure 7-2 shows a single classification error matrix for three vegetation/land cover categories (A, B, and C) and a change detection error matrix for the same three categories. The single classification matrix is of dimension 3 × 3, whereas the change detection error matrix is no longer of dimension 3 × 3 but rather 9 × 9. This is because we are no longer looking at a single classification but rather a change between two different classifications generated at different times. For both error matrices, one axis presents the three categories as derived from the remotely sensed

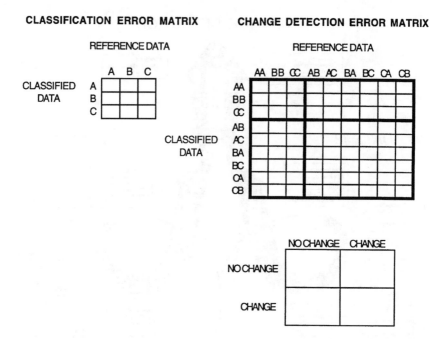

Figure 7-2 A comparison between a single classification error matrix and a change detection error matrix for the same vegetation/land use categories. Reproduced with permission, the American Society for Photogrammetry and Remote Sensing, from: Congalton, R.G. 1996. Accuracy assessment: A critical component of land cover mapping. IN: *Gap Analysis: A Landscape Approach to Biodiversity Planning.* A Peer-Reviewed Proceedings of the ASPRS/GAP Symposium. Charlotte, NC. pp. 119-131.

classification and the other axis shows the three categories identified from the reference data. The major diagonal of the matrices indicates correct classification. Off-diagonal elements in the matrices indicate the different types of confusion (called omission and commission error) that exist in the classification. This information is helpful in guiding the user to where the major problems exist in the classification.

When using the change detection error matrix the question of interest is, "What category was this area at time 1 and what is it at time 2?" The answer has nine possible outcomes for each dimension of the matrix (A at time 1 and A at time 2, A at time 1 and B at time 2, A at time 1 and C at time 2, ..., C at time 1 and C at time 2), all of which are indicated in the error matrix. It is then important to note what the remotely sensed data said about the change and compare it to what the reference data indicates. This comparison uses the exact same logic as for the single classification error matrix; it is just complicated by the two time periods (i.e., the change).

The change detection error matrix can also be simplified into a no-change/change error matrix. The no-change/change error matrix can be formulated by summing the cells in the four appropriate sections of the change detection error matrix (Figure 7-2). For example, to get the number of areas that both the classification and reference data correctly determined that no change had occurred between two

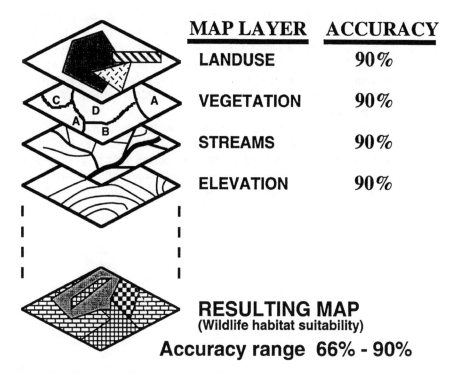

MAP LAYER	ACCURACY
LANDUSE	**90%**
VEGETATION	**90%**
STREAMS	**90%**
ELEVATION	**90%**

RESULTING MAP
(Wildlife habitat suitability)
Accuracy range 66% - 90%

Figure 7-3 The range of accuracies for a decision made from combining multiple layers of spatial data.

dates, you would simply add together all the areas in the upper left box (the areas that did not change in either the classification or reference data). You would proceed to the upper right box to find the areas that the classification detected no change and the reference data considered change. From the change detection error matrix and no-change/change error matrix, the analysts can easily determine if a low accuracy was due to a poor change detection technique, misclassification, or both.

Multilayer Assessments

Everything that has been presented in the book up to this point, with the exception of the last section on change detection, has dealt with the accuracy of a single map layer. However, it is important to at least mention multilayer assessments. Figure 7-3 demonstrates a scenario in which four different map layers are combined to produce a map of wildlife habitat suitability. In this scenario, accuracy assessments have been performed on each of the map layers and each layer is 90% accurate. The question is, how accurate is the wildlife suitability map?

If the four map layers are independent (i.e., the errors in each map are not correlated), then probability tells us that the accuracy would be computed by multiplying the accuracies of the layers together. Therefore, the accuracy of the final map is 90% × 90% × 90% × 90% = 66%. However, if the four map layers are not independent but rather completely correlated with each other (i.e., the errors are in the exact same place in all four layers), then the accuracy of the final map is 90%. In reality, neither of these cases are very likely. There is usually some correlation between the map layers. For instance, vegetation is certainly related to proximity to a stream and also to elevation. Therefore, the actual accuracy of the final map could only be determined by performing another accuracy assessment on this layer. We do know that this accuracy will be between 66% and 90%, and will probably be closer to 90% than to 66%.

One final observation should be mentioned here. It is quite eye-opening that using four map layers, all with very high accuracies, could result in a final map of only 66% accuracy. On the other hand, we have been using these types of maps for a long time without any knowledge of their accuracy. Certainly this knowledge can only help us to improve our ability to effectively use spatial data.

CHAPTER 8

The California Hardwood Rangeland Monitoring Project

INTRODUCTION

This chapter provides a detailed, real-world example of using the principles and practices outlined in this book to assess the accuracy of maps produced from both photo interpretation and the classification of digital satellite imagery. This specific case study was chosen for two reasons:

1. The assessment included analysis of the accuracy of maps created from both photo interpretation and satellite image classification, allowing for comparison of both mapping and accuracy assessment methods.
2. Numerous trade-offs between statistical rigor and practical implementation were required throughout the project.

As you will see, this case study is far from being the perfect example of accuracy assessment design, implementation, and analysis. The project was one of the first production accuracy assessments performed and, as such, offered ample opportunities for learning. Yet it is illustrative of problems typically encountered in accuracy assessment. The case study presents a real-world example with real world trade-offs and considerations. The implications of each decision are analyzed and discussed. The purpose of the case study is to make the reader fully aware of both the obvious and the subtle, yet critical considerations in designing and implementing an accuracy assessment.

BACKGROUND

Low use and low value have traditionally characterized California's hardwood rangeland resource. However, over the last 40 years increasing populations have forced development into hardwood rangelands, focusing new demands on hardwood lands, and resulting in changes in the extent and distribution of this resource. Hardwood stocking has declined, as has the number of acres of hardwoods with the conversion to industrial, residential, and intensive agricultural uses.

To assess and analyze the nature and implications of these changes, the California Department of Forestry and Fire Protection (CDF) instituted long-term monitoring of the hardwood resource as part of the Integrated Hardwood Range Management Program. In the late 1980s, the California Department of Forestry and Fire Protection contracted with the California Polytechnic Institute at San Luis Obispo to complete a map of the hardwood cover types in areas less than 5,000 feet in elevation within the State of California. This area is known as the hardwood rangeland zone. This photo-based map was derived from photo interpretation of 1981 aerial photography and portrays the type and extent of hardwood rangelands throughout the state (Pillsbury et al. 1991).

In late 1990, the CDF contracted with Pacific Meridian Resources to create a new map from satellite imagery and to assess the accuracy of both the new map and the photo-based map. This chapter concentrates on the methods, assumptions, and results of the accuracy assessment portions of the project. The sample design, data collection, and analysis methods used to assess the accuracy of both the photo and satellite derived maps are presented. Analysis results are discussed as well as the practical trade-offs apparent in each accuracy assessment task. Additional information of the methods used to create the maps can be found in *California Hardwood Rangeland Monitoring: Final Report* (Pacific Meridian Resources 1994).

The accuracies of four maps were assessed:

- Tree crown closure created from photo interpretation of 1981 aerial photography;
- Land cover type created from photo interpretation of 1981 aerial photography;
- Tree crown closure created from classification of 1990 digital satellite imagery;
- Land cover type created from classification of 1990 digital satellite imagery.

The organization of this chapter follows the organization of Chapters 3, 4, 5, and 6. First, the project's sample design is discussed. Next, data collection and methods are presented. Finally, the results of the accuracy analysis are detailed.

SAMPLE DESIGN

Sample design is critical to any accuracy assessment. The sample design for this project was extremely complex because it involved the assessment of four different maps (the 1981 photo and 1990 satellite maps) and used two types of reference data (the 1981 photos and field visits accomplished in 1991). As a result, trade-offs between statistical rigor and practicality are apparent throughout this case study. In particular, budget considerations directed the choice of source data. Because the state could not afford to fly new photography, existing aerial photography from 1981 was used as the primary source data for assessment of both the

1981 and 1990 maps. Use of the 1981 photos, in turn, drove much of the sample design, including the selection of the appropriate sample unit and the methods used to select the sample units.

Sample design for this project addressed three types of samples:

1. Samples from the 1981 map polygons for photo interpretation and assessment of both the 1981 photo and 1990 satellite maps.
2. Samples from the 1981 map polygons for field data collection and assessment of the 1981 photo-based map, the 1990 satellite-based map, and the 1992 photo interpretation of the 1981 photos.
3. Sample areas classified as hardwoods in the 1990 satellite-based map that fell outside of the extent of the 1981 photo-based map, to assess the accuracy of the extent of the 1981 map.

As with all accuracy assessments, sample design involved addressing the questions posed at the beginning of Chapter 3:

1. How is the map information distributed?
2. What is the appropriate sample unit?
3. How many samples should be taken?
4. How should the samples be chosen?

How Is the Map Information Distributed?

The study area is the hardwood rangeland of California, which forms a donut-shaped area around California's Central Valley, and is depicted in Figure 8-1. Almost all of California's hardwood tree and shrub species occur in the area. This project concentrates on the hardwood tree ecosystems.

The extent of the 1981 coverage was defined as areas where hardwood cover types occur in California below 5,000 feet in elevation. The extent of the 1990 coverage was initially defined to be that of the 1981 maps. However, while the 1990 maps were being produced, errors of omission were discovered in the 1981 maps. Accordingly, the extent of the 1990 maps was greatly expanded to include over 30 million acres of land. To assess possible errors of omission, accuracy assessment samples were taken in locations mapped as hardwoods on the 1990 map but omitted from the 1981 hardwood map.

The classification schemes for this project characterize California's hardwood rangelands by tree crown closure and land cover type (including hardwood cover types). Tree crown closure was classified into the following five classes:

1. 0% (non-hardwood)
2. 1–9%
3. 10–33%
4. 34–75%
5. 76–100%

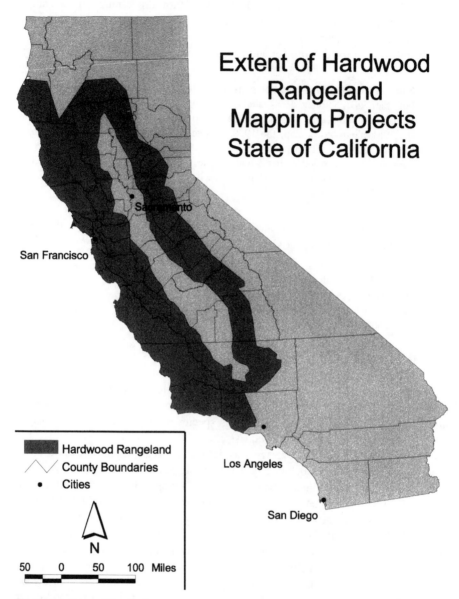

Figure 8-1 Map of the study area.

The land cover classification system consists of 12 classes:

1. Blue oak woodland
2. Blue oak/gray pine woodland*
3. Valley oak woodland

* Referred to as blue oak/digger pine woodland in the 1981 photo-based map. Digger pine is now called gray pine.

4. Coastal oak woodland
5. Montane hardwood
6. Potential hardwood
7. Conifer
8. Shrub
9. Grass
10. Urban
11. Water
12. Other

Figure 8-2 is a dichotomous key that illustrates the rules used to distinguish between land-cover type classes.

Like most land cover types, spatial autocorrelation exists in hardwood rangeland types. For example, California's annual summer drought results in hardwoods often being distributed in canyons and on northeast-facing slopes, where water stress is less than other areas.

What Is the Appropriate Sample Unit?

The preliminary accuracy assessment sampling design anticipated that the vegetation type polygons developed for the 1981 hardwood coverage could be used as the sampling units for accuracy assessment of both the 1981 and the 1990 maps. Unfortunately, use of this coverage as a source of sample units created multiple practical issues.

- First, using the 1981 polygons as the accuracy assessment sample units assumes that the polygons are homogeneous by crown closure and land cover type class, accurately delineated, and free of errors of omission. However, during the course of the project, significant errors of omission and polygon delineation were discovered in the existing maps. As a result many of the 1981 polygons had more class variation within the polygons than existed between the polygons. Many of the polygons exhibited (1) such highly variable crown closure and/or cover types that individual polygons actually consisted of two or more different crown closure classes and/or cover types and (2) arbitrary polygon boundaries through homogeneous vegetation types.
- Second, the existing polygon map also appeared to have been digitized at a much smaller scale than the photography, resulting in many straight edges that often extended beyond vegetation type rather than following the actual boundaries.
- Finally, many of the accuracy assessment polygons were several hundred acres in size, crossing several aerial photographs. Their large size made them impractical to photo interpret or to traverse in the field.

The following steps were taken to address these problems:

1. When the sample polygon contained multiple classes or was poorly delineated, a new homogeneous sample polygon was delineated *within* the original polygon. A box was delineated on the 1981 stereo photography within each randomly selected sample polygon. The box was placed inside an area of homogeneous crown closure

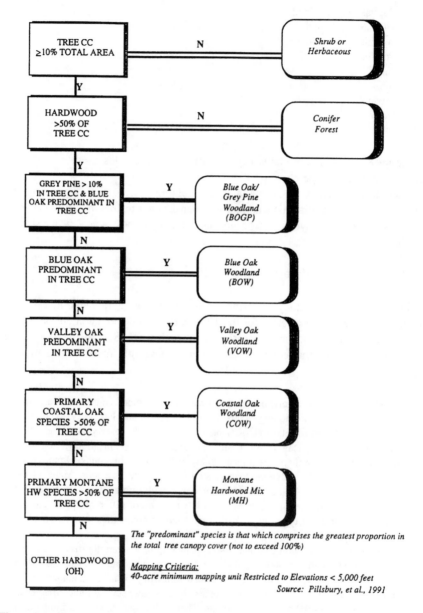

Figure 8-2 Decision tree for hardwood classification.

class and cover type. This site then became the center of a two or three sample cluster. Up to two additional boxes were delineated within an adjacent vegetation type differing in either density or cover type class.

2. When the sample polygons spanned more than one aerial photograph, a portion of the polygons existing on one aerial photograph was delineated as the final sample unit.

Table 8-1a   Sites Selected from 1981 Map by Site and Land Cover Type

Reference Cover-Type Label	Site Type[a] A	P	J	Total
Blue oak woodland	108	55	55	218
Blue oak/gray pine	16	5	20	41
Valley oak woodland	12	6	11	29
Coastal oak Woodland	71	72	45	188
Montane hardwood	112	30	29	171
Other hardwood	10	5	10	25
Non-hardwood	6	4	7	17
Total	335	177	177	689

[a] Site type key: A = sites photo interpreted in the office only; J = sites visited in the field; P = separate in-office photo interpretation of field-visited sites.

How Many Samples Should Be Taken?

A total of 817 accuracy assessment sites were sampled. Tables 8-1 a, b, and c summarize the number of accuracy assessment samples by sample type, cover type, and crown closure classes. Ideally, the 817 samples would have been allocated so that at least 50 samples would have been chosen from each crown closure or cover type class for both field and photo samples. As the tables show, while the goal of 50 samples was often met or exceeded for photo interpreted samples, it was not met for all classes or for field samples.

The reasons for these sampling deficiencies are varied and include the following practical considerations:

Table 8-1b Sites Selected from 1981 Map by Site and Tree Crown Closure Type

Reference Tree Crown Closure Label	Site Type[a]			
	A	P	J	Total
0%	6	4	8	18
1–9%	16	6	10	32
10–33%	124	57	54	235
34–75%	173	78	78	329
76–100%	16	32	27	75
Total	335	177	177	689

[a]Site type key: A = sites photo interpreted in the office only; J = sites visited in the field; P = separate in-office photo interpretation of field-visited sites.

- Sample polygon selection was essentially based on the vegetation distribution of the 1981 map. Inasmuch as the reference data has a different distribution than the map data (i.e., errors in the map exist), a potential exists for undersampling some classes. This also affects the sample distribution of the 1990 map, as the same sample polygons were used to assess it.
- The State of California is divided into regions for management and regulatory purposes. The California Department of Forestry requested that the contracted sample amount be distributed equally by region, and then by cover type within each region. Because all hardwood rangeland types do not occur in all regions, the prestratification of the samples caused some types to be undersampled.
- Valley oak (VOW) polygons are rare in the 1981 map, making it difficult to find enough polygons to sample.
- Field access was extremely difficult, making field data collection expensive. As a result, the budget dictated that compromises be made between travel cost and sample distribution.

Table 8-1c   Additional Sites Selected from 1990 Map by Site and Tree Crown Closure Type

Reference Tree Crown Closure Label	Site Type		
	A	J	Total
0%	21	11	32
1-9%	13	2	15
10-33%	18	9	27
34-75%	11	10	21
76-100%	25	8	33
Total	88	40	128

Site type key: A = sites photo interpreted in the office only; J = sites visited in the field; P = separate in office photo interpretation of field visited sites.

How Should the Samples Be Chosen and Distributed across the Landscape?

Despite suspected spatial autocorrelation in the distribution of hardwood range-lands, a cluster of sites was chosen for both photo-interpreted and field-visited sites. The choice of these sites was economically driven. Both photo interpretation set-up costs and field travel costs were greatly reduced by grouping samples together on one photo. Accuracy assessment samples were chosen using different procedures, depending on (1) if the reference data were to be field visits or photo interpretation and (2) if the sample unit was chosen from the 1981 coverage or from the 1990 coverage.

Samples Chosen from the 1981 Coverage

Both field and office samples were chosen from the 1981 coverage. Sites were selected for photo interpretation in the *office* using a random sample. Sampling was accomplished by

1. Stratifying the coverage into the hardwood cover types for each of the five California management regions.

2. Assigning a unique number to each of the polygons using the ARC/INFO PAT file.
3. Using a random number generator to select up to 20 polygons from each cover type that occurred in each region.
4. Using polygons from this sample population to select the center polygon from which two or three different sample sites would be selected and photo-interpreted in the office.

While this selection method was viable for office interpreted sites, random sampling could not be used to select accuracy assessment sites to be *field*-visited, because road-accessible sites could not be determined from the aerial photography. Five test trips to the field proved that more than 50% of randomly selected polygons lay along private ranch roads behind locked gates.

Accordingly, field-visited accuracy assessment sites were selected through a two-stage process. First, routes were chosen that both passed near or through many existing polygons and covered as much ecological variation within each image as possible. Site inaccessibility necessitated that field sample selection be partially dependent on the relative ease of access and observation.

Sites were selected for photo interpretation in the field in the following way. First, a 1:100,000 map was plotted of the image, polygons (without labels), and roads.

1. Field personnel determined which of the existing polygons were road-accessible by looking at the route delineated on the 1:100,000 scale draft classification maps that were used for field verification. To assure accessibility, routes were originally chosen wherever possible along public roads that intersected existing polygons. Any existing polygon that lay along this route could potentially be sampled for field verification.
2. To reduce potential site-selection bias, field personnel used dice to decide whether or not to sample an accessible polygon. Depending upon how many existing polygons were present in each ecoregion (subset of the imagery representing similar ecological conditions), a roll of one or more previously selected numbers (ranging from 1 to 6) on a single die indicated whether an existing polygon was to be sampled. If an ecoregion contained relatively more polygons than other ecoregions, two or three numbers might have been used on each roll of the die to select polygons. If an ecoregion contained relatively fewer polygons than other ecoregions, a single number may have been used. This was done to avoid over- or undersampling polygons within each ecoregion. For example, the number 6 may have been used to sample the relatively small ecoregion 46/32, while the numbers 1, 3, and 5 may have been used for the relatively large ecoregion 42/35-34. A minimum of 11 field sites per ecoregion were selected (11 sites × 15 ecoregions = 166 sites = one-third sample of 500 total accuracy sites).
3. A template was then used to delineate a box on the aerial photography within randomly selected roadside polygons. Up to two additional boxes were delineated within adjacent hardwood stands of either a different density or cover type class on the same photo.

Samples Chosen from the 1990 Coverage

Sample sites for testing the accuracy of the 1981 map's extent were selected by first randomly selecting 50 *potential hardwood* pixels per management region as *possible* sample points. More points were selected than would be needed because tests showed that many of the randomly selected points were located in areas outside the aerial photo coverage. Only those sites with available photo coverage could be used as accuracy assessment polygons. Using the *x,y* coordinates of each randomly selected pixel, a computer program generated a box around each pixel. Pixels were used because a polygon coverage of areas outside of the 1981 map did not exist.

Fifteen of the potential hardwood samples were selected per management region for assessment. A remote sensing analyst determined aerial photo availability for each potential polygon on a computer screen by zooming into each polygon location and displaying the polygon arcs and aerial photo flightlines over the TM imagery. The analyst started at the top of each image and worked down the image, checking each individual polygon; the first 15 polygons with photo coverage were transferred to the appropriate aerial photograph.

A total of 75 1990 map sites were finally selected. A subsample of 25 sites was selected for field verification. Field site selection was dependent upon the relative ease of access and observation. To take advantage of valuable field time, additional field sites were collected during field verification of office interpreted sites as a means of increasing sample size. The sites were randomly placed within hardwood stands larger than 20 acres adjacent to the randomly selected sites.

DATA COLLECTION

Once the complex sample design was complete, data collection was fairly straightforward because the same data were collected on all reference sites. As discussed in Chapter 4, data collection required addressing four basic questions:

1. What should be the source data for the reference samples?
2. What type of information should be collected for each sample?
3. When should the reference data be collected?
4. How do we ensure that the reference data are collected correctly, objectively, and consistently?

What Should Be the Source Data for the Reference Samples?

Both field visits and photo interpretation were used to collect reference information for the accuracy assessment reference sites. Budget constraints dictated use of the 1981 photography as the primary source data to assess the accuracy of both the 1981 and the 1990 maps. All sites were photo-interpreted in the office. Thus, the photo-interpreted 1981 map was assessed using the same photos as those used to create the map. Without assessing the accuracy of the photo interpretation, the result would have been more a comparison of two different photo interpretations than an accuracy assessment. Therefore, a subset of the photo interpreted sites was also field visited and additional field sites were taken.

Accuracy Assessment Photo Form

Site:_____-_____-____   HWPOLY-ID:_____   Date:___/___/___   Observer:_____
    type   region   #

Photo:_____-_____-_____   Photo Source:_____   Image:_____
    flightline   photo   quad

Observation Level: 1   2   3   4

% of total area occupied by tree cover

C#	Species	%

% of total area occupied by grass, bare, shrub, etc.

C#	Species	%

NOTE: Total tree and other cover observed on photo should equal 100%.

WHR Cover Type:_____   Size:_____

Current map polygon delineation: Very Poor   Poor   Good   Very Good

Describe delineation:_

C#	Comments

Figure 8-3 Accuracy assessment form.

What Type of Information Should Be Collected?

Both the 1981 and 1990 projects were concerned with mapping the extent, type, and condition of California hardwood rangeland. Each site was photo-interpreted in the office and/or the field, and an accuracy assessment form was completed that characterized the variation in land cover on the site (see Figure 8-3). Field personnel identified primary and associate hardwood cover type species by either driving through the site or doing a partial walk-through if accessibility allowed, or viewing the site from a distance through binoculars if the site was inaccessible.

For each site, the following were recorded:

1. Site—three-part alphanumeric accuracy assessment polygon label composed of the following codes:
   a. Type = A (photo-interpreted in office)
            J (photo-interpreted in the field)
            P (office photo interpretation of field-verified site)
   b. Region = SRPP region
   c. Number = sample number
2. HWPOLY-ID—item used to identify existing polygon in INFO, if available
3. Date—date of the photo interpretation
4. Observer—initials of the photo interpreter
5. Photo—photo number (identified by flightline, photo number, and quad, respectively) of aerial photograph on which accuracy assessment polygon has been delineated
6. Photo source—source agency for aerial photography (e.g., CDF, NASA, etc.) and photo job number if available
7. Image—identifies Landsat TM scene(s) polygon falls on using a four- or six-digit code indicating path/row(s) (e.g., 44/33, 44/32-33)
8. Observation level—used as an indicator of the potential accuracy of the photo interpretation, a "1" being the most accurate and a "4" being the least accurate:
   1. Walk through hardwood stand
   2. Viewing from road adjacent to hardwood stand
   3. Viewing from afar, i.e., road or ridge opposite hardwood stand
   4. Photo interpreted in office
9. Tree crown closure matrix—four-letter species codes used to record percentage of crown closure by primary and associate species (including gray pine) and "conifer"; includes numbered comments relating to species and crown closure calls in comment box.
10. Other cover crown closure matrix—four-letter species codes used to record percentage of crown closure occupied by the following non-tree cover types:
    a. Grass
    b. Shrub (if scrub oak, list percentage separate from other shrub)
    c. Urban
    d. Water
    e. Other (bare ground, agriculture, marsh, etc.)
11. WHR cover type—cover type calculated in the field or office using the Decision Tree for Mapping Hardwood Species Groups (Pillsbury et al. 1991) (see Figure 8-2) and recorded as follows:
    BOW = Blue oak woodland
    BOGP = Blue oak foothill/gray pine
    VOW = Valley oak woodland
    COW = Coastal oak woodland
    MH   = Montane hardwood
    OH   = Other hardwood
12. Size class—estimated average hardwood DBH recorded by size class:
    S < 12"
    L ≥ 12"
13. Current map polygon delineation—visual analysis of the general accuracy level of existing map polygon delineation as viewed on computer screen by overlaying polygons on the imagery using the following descriptions:

a. *Very poor:* existing polygon boundary does not follow hardwood stand along any of its perimeter; many unnatural contours; arbitrary polygon closure; polygon includes more than one density class, has a high level of variation in density or cover type, and has inclusions of non-hardwood cover or other hardwood cover types and densities within the 40 acre minimum mapping unit.

b. *Poor:* existing polygon boundary shifted away from actual hardwood stand perimeter; inclusions of non-hardwood cover or other hardwood cover types and densities within the 40 acre minimum mapping unit.

c. *Good:* existing polygon boundary generally follows hardwood stand perimeter; no inclusions of non-hardwood cover or other hardwood cover types or densities within the 40 acre minimum mapping unit.

d. *Very Good:* existing polygon boundary tightly follows hardwood stand along entire perimeter; inclusions of non-hardwood cover within the 40 acres are delineated; hardwood stand has evenly distributed crown closure and homogeneous cover type throughout polygon.

14. Cover type and density fuzzy logic matrix—each polygon evaluated for the likelihood of being identified as each of the six possible cover types and four possible existing crown closure classes. "Likelihood" is indicated using the terms "absolutely wrong," "probably wrong," "acceptable," "probably right," and "absolutely right" (Figure 8-4).

Following completion of the forms, all data were entered into a database for later analysis. In addition, upon completion of the field data collection, field-verified accuracy assessment site boundaries were captured using heads-up digitizing.

When Should the Reference Data Be Collected?

Unfortunately, the only aerial photos available for the assessment were the 1981 1:24,000 panchromatic photography used to develop the 1981 map. While forest land typically does not change as quickly as agricultural or urban land, the 9-year difference between the 1981 photos and the 1990 imagery caused problems in the assessment. Fires, harvesting, and urban development changed several accuracy assessment sites between the date of the photos (1981), the date of the imagery (1990), and the date of the field visits (1992). Only those sites that had not changed significantly in the field were included in the field sample. However, it is impossible to know how many of the sites that were not field-visited also changed between 1981 and the date of the imagery.

Quality Control

Data Independence. Because they were completed by two different organizations, the assessment of the 1981 map was completely independent of the effort to create the 1981 map. In addition, independence was also imposed in the assessment of the 1990 map. Accuracy assessment data were always kept separate from any information used to make the 1990 map. At no time did the accuracy assessment photo interpreters have any knowledge of the map labels for either the 1981 or 1990 map.

Data Consistency. Data consistency was imposed in several ways. First, an accuracy assessment manual was developed which clearly explained all data collection procedures. Second, as illustrated in Figures 8-3 and 8-4, personnel used forms to collect all accuracy assessment data. Finally, all personnel were trained simultaneously and the project manager frequently reviewed their work.

Cover Type/ Density	Absolutely Right	Probably Right	Acceptable	Probably Wrong	Absolutely Wrong
Blue Oak					
Blue Oak/ Grey Pine					
Valley Oak Woodland					
Coastal Oak Woodland					
Montane Hardwood					
Other					
1 (<10%)					
2 (10-33%)					
3 (34-75%)					
4 (76-100%)					

Figure 8-4 Cover type and density fuzzy logic matrix form.

Data Quality. The map location of accuracy assessment sites was derived directly from the 1981 map because the sites were chosen from the 1981 map polygons. Location of the site on the reference data (the 1981 photos) was accomplished by viewing the sample polygon's boundaries over the satellite imagery, and then transferring the site location onto the photo by matching flightline location, roads, streams, and patterns of vegetation.

To minimize photo interpretation error, personnel most familiar with the vegetation in each region interpreted the sample polygons. Species identification in the office was enhanced through the use of ancillary data, including extensive field notes and ecological information concerning the distribution of hardwood types (Griffin and Critchfield, 1972).

Data entry was done once. To check the quality of the entry, sample database fields were compared to the original information on the forms. However, data entry was not perfect and caused later problems in analysis of the error matrices.

ANALYSIS

Development of the Error Matrices

The first step in accuracy assessment analysis requires the development of error matrices. Error matrices, in turn, require labeling the samples. As introduced in Chapter 2, each accuracy assessment site in an error matrix has two labels:

1. *The reference site label* refers to the label derived from data collected either from field or office photo interpretation that makes up the reference data (the data against which the map is compared) during accuracy assessment.
2. *The map site label* refers to the map label of the accuracy assessment site. In this project the map label is derived either from the existing 1981 photo-interpreted map or from decision rules applied to the pixel composition of the site on the 1990 satellite map.

Reference labels were calculated both with (1) deterministic labels that automate the classification systems, and (2) labels that account for variation in interpretation. Deterministic labels were calculated for each sample site from the percent crown closure estimates. Calculated *crown closure* labels were determined using the classification system rules presented earlier in this chapter. Calculated *cover type* labels were determined using the Decision Tree for Mapping Hardwood Species Groups (Figure 8-2) provided by the California Department of Forestry and Fire Protection. Unfortunately, this classification scheme is not totally exhaustive, resulting in several sample sites receiving no cover type label.

Variance labels also were created to (1) account for variation in estimates, and (2) deal with the imprecision in the cover type classification system. For this project, both expert and measured approaches to fuzzy set theory in accuracy assessment were implemented. The measured approach measures the variance from paired interpretations of the same site and removes that variance from the difference matrix. Two independent interpretations exist for each accuracy assessment reference site that was photo-interpreted both in the field and in the office. Because the site was held constant while the interpreter varied, these pairs of interpretations can be used to measure variation in interpretation. This method is fairly simple to implement with vegetation class characteristics such as crown closure which are represented by discrete breaks in a continuum on *one* variable. The algorithms for implementing this method on class characteristics represented by discrete breaks in *multiple* variables (e.g., cover type as a function of percent crown closure of several hardwood species types) are less defined and more difficult to implement. For this reason the methods used by Gopal and Woodcock (1994) were implemented for the labeling of cover type reference sites.

Map accuracy site labels for the 1981 map were taken directly from the map label for that site. *Map* labels for the 1990 map were calculated for each site using (1) the site's pixel composition of crown closure and cover type raster data layers, and (2) algorithms based on the classification system's decision rules. Thus, a polygon could receive a label that was the result of the mixture of pixels in the polygon. For example, an accuracy assessment sample polygon comprised of a mixture of only closed canopy (76–100%) and open pixels (1–10%) would receive a crown cover label that was the average of the pixel values (perhaps 35–75%).

Once the labels were created, the error matrices were built. Tables 8-2 a, b, c, and d show the initial error matrices for the four maps assessed. As with most accuracy assessments, the first matrices are far from being the final matrices. In fact, it is probably more correct to name the initial matrices as *difference* matrices because they indicate that differences (and not necessarily map errors) exist between the reference and map labels.

Table 8-2a Crown Closure Difference Matrix, 1981 Photo-Interpreted Map

Reference Data

Class	0%	1-9%	10-33%	34-75%	76-100%	Total
0%	0	1	9	8	1	19
1-9%	4	17	78	34	3	136
10-33%	1	2	59	69	7	138
34-75%	2	1	21	93	11	128
76-100%	3	0	2	37	17	59
Total	10	21	169	241	39	480

Existing Map

Producer's Accuracy

Reference	Percent
0%	0
1-9%	81
10-33%	35
34-75%	39
76-100%	44

User's Accuracy

Map	Percent
0%	0
1-9%	13
10-33%	43
34-75%	73
76-100%	29

OVERALL ACCURACY = 186/480 = 39%

Table 8-2b   Crown Closure Difference Matrix, 1990 Satellite Interpreted Map

Reference Data

	Class	0%	1-9%	10-33%	34-75%	76-100%	Total
Updated Map	0%	4	1	1	7	5	18
	1-9%	1	11	42	12	0	66
	10-33%	1	7	97	84	4	193
	34-75%	4	2	29	135	22	192
	76-100%	0	0	0	3	8	11
	Total	10	21	169	241	39	480

Producer's Accuracy

Reference	Percent
0%	40
1-9%	52
10-33%	57
34-75%	56
76-100%	21

User's Accuracy

Map	Percent
0%	22
1-9%	17
10-33%	50
34-75%	70
76-100%	73

OVERALL ACCURACY = 255/480 =53%

Table 8-2c   Cover Type Difference Matrix, 1981 Photo-Interpreted Map

	Class	NH	BOGP	BOW	COW	MH	VOW	Total
				Reference Data				
Existing Map	NH	0	1	8	5	4	2	20
	BOGP	1	19	40	4	22	5	91
	BOW	2	6	68	4	12	7	99
	COW	2	1	22	54	10	5	94
	MH	3	5	8	27	81	1	125
	VOW	1	0	7	14	8	2	32
	Total	9	32	153	108	137	22	461

Producer's Accuracy

Reference	Percent
NH	0
BOGP	59
BOW	25
COW	5
MH	59
VOW	9

User's Accuracy

Map	Percent
NH	0
BOGP	21
BOW	69
COW	63
MH	65
VOW	6

OVERALL ACCURACY = 224/461 = 49%

Table 8-2d   Cover Type Difference Matrix, 1990 Satellite-Interpreted Map

		Reference Data						
Class	NH	BOGP	BOW	COW	MH	VOW	Total
NH	4	0	1	2	9	0	16
BOGP	1	18	27	3	18	3	70
BOW	0	9	98	7	18	10	142
COW	2	0	12	71	9	6	100
MH	1	5	8	20	75	1	110
VOW	1	0	7	5	8	2	23
Total	9	32	153	108	137	22	461

Updated Map

Producer's Accuracy

Reference	Percent
NH	44
BOGP	56
BOW	64
COW	66
MH	55
VOW	9

User's Accuracy

Map	Percent
NH	25
BOGP	26
BOW	69
COW	71
MH	68
VOW	9

OVERALL ACCURACY = 268/461 = 58%

Two types of analysis should be carried out on the error matrices. First, we must determine if the results in the matrix are statistically valid (see Chapter 5). Next, we need to learn what causes samples to fall off the diagonal (see Chapter 6).

Statistical Analysis

Statistical analyses, including normalizing the matrices using the iterative proportional fitting procedure (i.e., Margfit) and the Kappa measure of agreement, were performed on these difference matrices. The normalization process allows for individual cell values within matrices to be directly compared without regard for sample size differences. A normalized accuracy was computed for each matrix. Table 8-3 a–d present the results of the normalization.

The results of the Kappa analysis are shown in Tables 8-4 and 8-5. A test of significance of an individual matrix was performed to see if the classification process was significantly better than a random assignment of pixels. Table 8-4 shows that these results were significant for all four matrices. Table 8-5 presents the results of the appropriate pairwise comparisons. This test determines if the difference between two error matrices is statistically significant. In this example, it was appropriate to compare the results of the crown closure maps generated from 1981 aerial photo interpretation and 1990 satellite image processing. It was also appropriate to compare the cover type map derived from the 1981 aerial photo interpretation with the 1990 cover type map created from satellite image processing. In both of these cases, the matrices (and therefore, the maps) were significantly different from each other. By examining the accuracy measures, it could be concluded that the 1990 maps generated from satellite imagery were significantly better than the maps created in 1981 from aerial photography.

Analysis of Off-Diagonal Samples

Following the statistical analysis of the matrices, the off-diagonal elements of the matrix need to be examined for possible

1. Errors in the reference data
2. Sensitivity of the classification schemes to observer variability
3. Inappropriateness of the photo interpretation or satellite remote sensing for mapping hardwood rangeland crown closure and cover type, and
4. Mapping error.

Crown Closure Analysis

Assessing errors in crown closure is extremely difficult because crown closure is rarely measured. Therefore, it is difficult to analyze the possibility of errors in crown closure reference data. To learn if the causes of differences in the matrix resulted from error or from variation in interpretation, two independent photo interpretations of the same accuracy assessment reference site were made for 173 sites: one in the office and one in the field. No two pairs of interpretations were made by the same photo interpreter. Table 8-6 compares these interpretations. In general, the

Table 8-3a Normalized Crown Closure Difference Matrix, 1981 Photo-Interpreted Map

Reference Data

	Class	0%	1-9%	10-33%	34-75%	76-100%
	0%	0.1672	0.3171	0.2460	0.1383	0.1319
Existing Map	1-9%	0.1858	0.4568	0.2510	0.0693	0.0380
	10-33%	0.1150	0.1212	0.3533	0.2593	0.1512
	34-75%	0.1969	0.0747	0.1312	0.3584	0.2382
	76-100%	0.3351	0.0303	0.0185	0.1747	0.4407

Normalized Accuracy = 36%

Table 8-3b Normalized Crown Closure Difference Matrix, 1990 Satellite-Interpreted Map

Reference Data

	Class	0%	1-9%	10-33%	34-75%	76-100%
	0%	0.5670	0.1446	0.0276	0.0948	0.1665
Updated Map	1-9%	0.0838	0.4919	0.3470	0.0701	0.0067
	10-33%	0.0483	0.1848	0.4586	0.2729	0.0348
	34-75%	0.1514	0.0644	0.1450	0.4573	0.1819
	76-100%	0.1494	0.1143	0.0218	0.1049	0.6101

Normalized Accuracy = 52%

Table 8-3c   Normalized Cover Type Difference Matrix, 1981 Photo-Interpreted Map

Reference Data

	Class	NH	BOGP	BOW	COW	MH	VOW
	NH	0.1273	0.1619	0.1790	0.1418	0.1160	0.2740
	BOGP	0.0824	0.4539	0.1838	0.0250	0.1251	0.1300
Existing Map	BOW	0.1576	0.1737	0.3570	0.0287	0.0798	0.2034
	COW	0.1792	0.0456	0.1334	0.3957	0.0762	0.1697
	MH	0.1921	0.1280	0.0386	0.1529	0.4529	0.0354
	VOW	0.2614	0.0369	0.1081	0.2559	0.1500	0.1875

NORMALIZED ACCURACY = 33%

Table 8-3d   Normalized Cover Type Difference Matrix, 1990 Satellite-Interpreted Map

Reference Data

	Class	NH	BOGP	BOW	COW	MH	VOW
	NH	0.5963	0.0522	0.0373	0.0699	0.1824	0.0628
	BOGP	0.0536	0.5212	0.1844	0.0264	0.0959	0.1187
Updated Map	BOW	0.0123	0.1840	0.4539	0.0389	0.0659	0.2447
	COW	0.0896	0.0141	0.0840	0.5412	0.0494	0.2209
	MH	0.0622	0.1797	0.0661	0.1795	0.4537	0.0590
	VOW	0.1860	0.0489	0.1744	0.1440	0.1528	0.2939

NORMALIZED ACCURACY = 48%

Table 8-4 Individual Error Matrix Kappa Analysis Results

Error Matrix	KHAT	Variance	Z statistic
Table 8-2a	0.17	0.0010371	5.4
Table 8-2b	0.28	0.0011688	8.1
Table 8-2c	0.34	0.0008001	12.1
Table 8-2d	0.45	0.0008323	15.6

Table 8-5 Kappa Analysis Results for the Pairwise Comparison of the Error Matrices

Pairwise Comparison:	Z statistic
Table 8-2a vs. Table 8-2b	2.2036
Table 8-2c vs. Table 8-2d	2.6703

class values of the paired interpretations fall along the range of the diagonal, clearly illustrating the impacts of variation in interpretation.

The average difference in crown closure estimates between the office photo-interpreted and field photo-interpreted estimates was 9.31% with a standard deviation of 10.85%. To compensate for the impacts of variation in human interpretation on map accuracy assessment, a ±9% variance in crown closure was implemented on all office-interpreted sites. For example, a field-interpreted estimate of 11% crown closure would be considered comparable to a photo-interpreted estimate of either 1–9% class (i.e., 11 – 9 = 2) or the 10–33% class (i.e., 11 + 9 = 20). Table 8-7 illustrates how the variances were implemented across all crown closure classes.

Table 8-8 illustrates the implementation of the ranges on the matrix comparing the pairs of sites. A total of 16 sites fall outside of the allowable ranges. The photo interpretation of these 16 sites differs because of

- *Photo interpretation error.* At two sites the office photo interpreters mislabeled hardwoods for shrub.
- *Sensitivity of the classification system to observer variability.* One site differs in its labels of hardwood versus non-hardwood because the site is mixed hardwood/conifer. A 9% variance on the estimates of hardwood or conifer would place the site in a different category. The photo interpreter in the field noted that a hardwood classification would have been acceptable. The remaining 13 sites represented

Table 8-6 Comparison of Office and Field Photo Interpretation of Crown Closure

		Field Photo Sites				
Class	0%	<10%	10-33%	34-75%	76-100%	Total
Office Photo Sites 0%	4	0	0	0	0	4
1-9%	0	3	3	0	0	6
10-33%	1	4	41	11	0	57
34-75%	1	1	10	60	3	75
76-100%	1	0	0	7	23	31
Total	7	8	54	78	26	173

Producer's Accuracy

Reference	Percent
0%	57
1-9%	38
10-33%	76
34-75%	77
76-100%	88

User's Accuracy

Map	Percent
0%	100
1-9%	50
10-33%	72
34-75%	80
76-100%	74

OVERALL ACCURACY = 131/173 = 76%

Table 8-7   Crown Cover and Corresponding Acceptable Labels

If Field Site Equals:	Acceptable Photo Labels
0%	0 or 1-9
1-9%	0 or 1-9 or 10-33
10-18%	1-9 or 10-33
19-24%	11-33
25-42%	11-33 or 34-75
43-66%	34-75
67-84%	34-75 or 76-100
85-100%	76-100

variance in the photo interpretation estimates that are beyond those allowed in the adopted ranges. The 9% variance is an *average* value. By adopting the average (rather than the complete measured spread of variation), we are accepting that some of the differences in the matrices will be counted as map errors when they really are differences caused by variation in human interpretation.

The ±9% variance in crown closure estimates brings into question the appropriateness of photo interpretation for labeling crown closure. Will the labels be acceptable given that each label could vary by as much as 9% in either direction? The answer to this question lies in the anticipated uses of the map. For over 50 years, land managers and regulators have accepted photo interpretation for crown closure with few investigations concerning the accuracy of the photo-interpreted maps. This history of acceptance indicates that the relative nature of crown closure estimates is "good enough" for many applications. However, it is important to be aware of the variance of crown closure estimates whether they are used to create a map or to assess another map.

Crown Closure Map Results

Table 8-9 presents the crown closure error matrices for the 1981 map. The matrix includes the ±9% variance, resulting in an increase in overall accuracy from the 39% in Table 8-2a to 60% in Table 8-9. In general, the 1981 map systematically underestimates crown closure. In addition, errors of omission are significant with 18 of the 19 map non-hardwood sites actually containing more than 10% crown closure in hardwoods. Most of these errors of omission resulted from the hardwoods being misidentified as shrubs. Errors of commission of non-hardwoods to hardwoods were few and almost always included mixed hardwood–conifer stands where the map photo interpreter estimated there was more conifer than hardwood on the site.

Table 8-10 presents the crown closure error matrix for the 1990 map. Overall agreement between the reference data and the map is 73%, an increase from 53% in Table 8-2b. Like the 1981 maps, there seems to be a systematic underestimation of crown closure. However, unlike the 1981 maps, the updated maps suffer from the 10-year difference in between the date of aerial photography used for reference data and the date of the imagery.

Table 8-8 Comparison of Office and Field Photo Interpretation Adjusted for Variation in Crown Closure

Field Photo Sites

	Class	0%	1-9%	10-18%	19-24%	25-42%	43-66%	67-84%	85-100%	Total
Office Photo Sites	0%	4	0	0	0	0	0	0	0	4
	1-9%	0	3	2	0	0	1	0	0	6
	10-33%	1	4	12	10	24	5	1	0	57
	34-75%	1	1	1	0	27	31	14	0	75
	76-100%	1	0	0	0	0	4	6	20	31
	Total	7	8	15	10	51	40	21	20	172

Producer's Accuracy

Reference	Percent
0%	57
1-9%	88
10-18%	93
19-24%	100
25-42%	98
43-66%	78
67-84%	95
85-100%	100

User's Accuracy

Map	Percent
0%	100
1-9%	83
10-33%	88
34-75%	96
76-100%	84

OVERALL ACCURACY = 157/172 =91%

Table 8-9 Crown Closure Error Matrix, 1981 Photo-Interpreted Map

Reference Data

	Class	0%	1-9%	10-18%	19-24%	25-42%	43-66%	67-84%	85-100%	Total
	0%	0	1	2	2	8	4	1	1	19
Existing	1-9%	4	17	40	20	33	15	5	2	136
Map	10-33%	1	2	18	13	59	32	10	3	138
	34-75%	2	1	4	4	36	54	18	9	128
	76-100%	3	0	0	1	11	20	13	11	59
	Total	10	21	64	40	147	125	47	26	480

Producer's Accuracy

Reference	Percent
0%	40
1-9%	95
10-18%	91
19-24%	33
25-42%	65
43-66%	43
67-84%	66
85-100%	42

User's Accuracy

Map	Percent
0%	5
1-9%	45
10-33%	67
34-75%	84
76-100%	41

OVERALL ACCURACY = 286/480 = 60%

Table 8-10 Crown Closure Error Matrix, 1990 Satellite-Interpreted Map

Reference Data

	Class	0%	1-9%	10-18%	19-24%	25-42%	43-66%	67-84%	85-100%	Total
Updated Map	0%	4	1	0	1	0	4	4	4	18
	1-9%	1	11	21	7	20	5	1	0	66
	10-33%	1	7	36	24	79	35	10	1	193
	34-75%	4	2	7	8	47	79	32	13	192
	76-100%	0	0	0	0	1	2	0	8	11
	Total	10	21	64	40	147	125	47	26	480

Producer's Accuracy

Reference	Percent
0	50
1-9%	90
10-18%	89
19-24%	60
25-42%	86
43-66%	63
67-84%	68
85-100%	31

User's Accuracy

Map	Percent
0	28
1-9%	50
10-33%	76
34-75%	82
76-100%	73

OVERALL ACCURACY = 350/480 = 73%

Reductions in crown closure caused by harvesting, fires, and urban expansion will cause non-map error differences between the office photo interpretation of the 1981 photos and the 1990 maps.

Furthermore, it is assumed that raster map polygon labeling procedures produce the "true" label for a particular site. It is possible, however, to develop several different labeling algorithms that produce several different labels. The present labeling algorithm assumes that the midpoint of a class is a good estimate for class value. Use of the average percentage crown closure for each class increases accuracy by approximately 2%.

Errors of omission also exist on the updated maps. Nine of 13 samples mislabeled as non-hardwoods in the 1990 satellite map were labeled as hardwoods in the existing 1981 photo interpreted map. Like the existing map, errors of commission occur in mixed hardwood–conifer sites.

Cover Type Analysis

A Guide to Wildlife Habitats of California (Mayer and Laudenslayer, 1988) presents an excellent example of the absence of precise class boundaries in the vegetation classification systems for California. For example, the composition of coastal oak woodland and montane hardwood are often indistinguishable as follows:

> *Montane hardwood:* "In the Coast range ... Knobcone pine, Digger pine, Oregon white oak, and coast live oak are abundant at lower elevations" (p. 72).
> *Coastal oak woodland:* "In the North Coast Range to Sonoma County, Oregon white oak is the common deciduous oak ... Digger pine is common" (p. 78).

In an effort to remove some of the ambiguity in the system, the dichotomous key of Figure 8-2 was developed for this project. However, the key is based on estimates of crown closure by species type which are highly susceptible to variation in interpretation.

To understand the sensitivity of the matrix to variation in crown cover estimates versus errors in photo interpretation, the 173 sites used above in the crown closure analysis were also interpreted for cover type. Table 8-11 shows the results of comparing the photo-interpreted labels against one another. The off-diagonal elements are caused either by photo interpretation error (e.g., the misidentification of species) or by variation in interpretation. For example, of the 10 sites showing differences between BOW and BOGP labels, 9 were recognized by the field photo interpreter as acceptable as either BOW or BOGP. All eight sites contained gray pine. At issue was the crown closure percentage of gray pine.

To account for the variation, interpreters filled out a cover type fuzzy logic matrix for every accuracy assessment site. Each polygon was evaluated for the likelihood of being identified as each of the six possible cover types (Gopal and Woodcock, 1992). "Likelihood" was measured using the terms "absolutely wrong," "probably wrong," "acceptable," "probably right," and "absolutely right." For example, if a site with 65% total tree crown closure consisted of 20% valley oak, 25% blue oak, 10% gray pine, and 10% interior live oak, then the following interpretation might occur:

Table 8-11 Comparison of Office versus Field Photo Interpretation of Cover Type

| | | Field Photo Sites | | | | | |
Class	NH	BOGP	BOW	COW	MH	VOW	Total
NH	4	0	0	0	0	0	4
BOGP	0	3	1	0	1	0	5
BOW	1	9	38	0	4	2	54
COW	0	2	10	41	7	5	65
MH	1	4	4	1	14	1	25
VOW	0	0	1	0	1	3	5
Total	6	18	54	42	27	11	158

Office Photo Sites

Producer's Accuracy

Reference	Percent
NH	67
BOGP	17
BOW	70
COW	98
MH	52
VOW	27

User's Accuracy

Map	Percent
NH	100
BOGP	60
BOW	70
COW	63
MH	56
VOW	60

OVERALL ACCURACY = 103/158 = 65%

Blue oak woodland   → acceptable
Blue oak gray pine   → probably right
Valley oak woodland   → acceptable
Coastal oak woodland   → probably wrong
Montane hardwood   → probably wrong

A site with 100% grass cover, on the other hand, would be interpreted as follows:

Blue oak woodland   → absolutely wrong
Blue oak gray pine   → absolutely wrong
Valley oak woodland   → absolutely wrong
Coastal oak woodland   → absolutely wrong
Montane hardwood   → absolutely wrong

Table 8-12 incorporates the fuzzy logic "acceptable" interpretations into the confusion matrix. The greater the heterogeneity of the site, the more likely it is to have more than one "acceptable" label. The increase in overall agreement from 65%–80% shows that 15% of the disagreement was caused by variation in interpretation rather than photo interpreter error.

Sites where differences persist in the matrix occur primarily because of *photo interpretation error.* Hardwood species can be extremely difficult to distinguish in the field, much less from aerial photography.

- Almost all of the differences between BOW/BOGP and COW were caused by the office photo interpreter misidentifying coast live oak as blue oak.
- The differences between BOW/BOGP and MH are caused by misidentification of coast live oak as interior live oak, and by an artifact of the dichotomous key that mistakenly allows mixed interior live oak–blue oak stands to be labeled montane hardwood.
- Differences between VOW and BOW or COW occur because of misidentification of valley oak as either blue oak or coast live oak.

Cover Type Map Results

Tables 8-13 and 8-14 present the error matrices for cover type for the 1981 photo-interpreted and 1990 satellite-interpreted maps. Both tables include all "right," "probably right," and "acceptable" labels as matches on the diagonal of the matrix. Increases in overall accuracies of 7–9% over Tables 8-2 c and d indicate the impact of variation in interpretation on the matrices. As with crown closure, ambiguity exists between class labels for sites on the margins of class boundaries.

Overall accuracies of the 1990 map exceed that of the 1981 map. Incorporation of ancillary data (including the 1981 maps) and editing from field notes significantly increased the 1990 map's cover type accuracy. Specifically, much of the 1981 confusion between BOGP and BOW was reduced, as was that between COW and BOW/BOGP. However, the results do *not* constitute a comparison of photo

Table 8-12 Comparison of Office versus Field Photo Interpretation of Cover Type Using the Fuzzy Logic Approach

				Office Photo Sites				
	Class	NH	BOGP	BOW	COW	MH	VOW	Total
Field Photo Sites	NH	4	0	0	0	0	0	4
	BOGP	0	4	1	0	0	0	5
	BOW	0	0	49	0	3	2	54
	COW	0	2	9	48	2	4	65
	MH	0	2	2	1	19	1	25
	VOW	0	0	1	0	1	3	5
	Total	4	8	62	49	25	10	158

Producer's Accuracy

Reference	Percent
NH	100
BOGP	80
BOW	91
COW	74
MH	76
VOW	60

User's Accuracy

Reference	Percent
NH	100
BOGP	50
BOW	79
COW	98
MH	76
VOW	30

OVERALL ACCURACY = 127/158 = 80%

Table 8-13 Cover Type Error Matrix, 1981 Photo-Interpreted Map

Reference Data

Existing Map

Class	NH	BOGP	BOW	COW	MH	VOW	Total
NH	0	1	8	5	4	2	20
BOGP	1	30	33	4	19	4	91
BOW	2	1	77	5	9	5	99
COW	2	0	22	59	9	2	94
MH	2	3	8	23	88	1	125
VOW	1	0	7	14	7	3	32
Total	8	35	155	110	136	17	461

Producer's Accuracy

Reference	Percent
NH	0
BOGP	86
BOW	50
COW	54
MH	65
VOW	18

User's Accuracy

Map	Percent
NH	0
BOGP	33
BOW	78
COW	63
MH	70
VOW	9

OVERALL ACCURACY = 257/461 = 56%

Table 8-14   Cover Type Error Matrix, 1990 Satellite-Interpreted Map

			Reference Data				
Class	NH	BOGP	BOW	COW	MH	VOW	Total
NH	4	0	1	2	9	0	16
BOGP	1	31	17	3	15	3	70
BOW	0	1	111	7	16	7	142
COW	2	0	11	77	7	3	100
MH	0	3	8	16	82	1	110
VOW	1	0	7	5	7	3	23
Total	8	35	155	110	136	17	461

Updated Map

Producer's Accuracy

Reference	Percent
NH	50
BOGP	89
BOW	72
COW	70
MH	60
VOW	18

User's Accuracy

Map	Percent
NH	25
BOGP	44
BOW	78
COW	77
MH	75
VOW	13

OVERALL ACCURACY = 308/461 = 67%

interpretation versus satellite image processing methods, because the 1981 map was an extremely important ancillary layer in the creation of the 1990 map.

The most significant confusion in both maps' error matrices persists between BOW and BOGP, BOGP/BOW and MH, MH and COW, and VOW and all other hardwood types:

- Differences caused by various estimates of the amounts of gray pine in blue oak stands continue to cause confusion between BOGP and BOW labels. Twelve of the 18 BOW–BOGP confused sites contained various percentages of gray pine. Because of its sparse canopy, gray pine is extremely difficult to see and estimate on both aerial photography and satellite imagery.
- Error in species identification continues to contribute to the differences. As Table 8-12 indicates, this confusion also occurs in the reference data and is contributing to lower map accuracies. It is difficult to distinguish the live oaks from one another on the ground, much less from remotely sensed data. Similarly, valley oak on slopes is difficult to distinguish from blue oaks. Valley oak woodland is the most confused type. It is unlikely that mixed valley oak stands on slopes can be adequately classified from either aerial photography or satellite imagery. Conversely, the confusion between blue oak and the live oaks could be aided through the use of multitemporal imagery or photography, because blue oak is deciduous whereas the live oaks are evergreen.
- Much of the apparent confusion may be a function of failures in the classification system. The system does not adequately address the classification of hardwood types that are dominated by interior live oak or Oregon white oak. For example, of the sites identified by the reference data as MH when the maps listed BOW or BOGP, 20 of the 31 sites were pure interior live oak or mixtures of only interior live oak and blue oak with some sites containing gray pine. At present the classification system labels such stands as MH, yet their composition would more support a label of BOW or BOGP as described in *A Guide to Wildlife Habitats of California* (Mayer and Laudenslayer, 1988).
- Most MH indicator species—including interior live oak, black oak, tan oak, Pacific madrone, and canyon live oak—in addition to many associate species—particularly California bay laurel, valley oak, and coast live oak itself—also commonly occur in COW stands. This is further compounded by the fact that taxonomic similarities between coast live oak and interior live oak are very difficult to differentiate in the field, much less on aerial photography or satellite data.

Extent

The final error matrices assess the accuracy of the extent of the 1981 and 1990 maps. As Tables 8-15 and 8-16 show, the 1981 maps contain significant errors of omission. Of the 102 accuracy assessment sites labeled non-hardwood on the 1981 map but hardwood on the 1990 map, 86 (84%) were labeled as hardwood by the reference data. The extent of the 1990 maps is more accurate than that of the 1981 maps. This is especially significant given the fact that no editing or quality control was performed on the 1990 maps in areas outside of the extent of the 1981 map.

Table 8-15 Error Matrix of Extent, 1981 Photo-Interpreted Map

Reference Data

	Class	non-hardwood	hardwood	Total
1981 Map	non-hardwood	16	86	102
	hardwood	10	451	461
	Total	26	537	563

Producer's Accuracy		User's Accuracy	
Reference	Percent	Map	Percent
Non-Hdwd	62	Non-Hdwd	16
Hardwood	84	Hardwood	98

OVERALL ACCURACY = 467/563 = 83%

Table 8-16 Error Matrix of Extent, 1990 Satellite-Interpreted Map

	Class	non-hardwood	hardwood	Total
Updated Map	non-hardwood	5	16	21
	hardwood	21	521	542
	Total	26	537	563

Producer's Accuracy		User's Accuracy	
Reference	Percent	Map	Percent
Non-Hdwd	19	Non-Hdwd	24
Hardwood	97	Hardwood	96

OVERALL ACCURACY = 526/563 = 93%

DISCUSSION

The 1990 mapping project resulted in the creation of GIS land cover information for over 32 million acres of land; approximately one third of the State of California and three times the extent of the 1981 mapping project. Of this 32 million acres,

- 11.4 million are hardwood rangelands,
- 3.8 million are conifer lands,
- 3.9 million are shrublands,
- At least 7.9 million are grasslands,
- 0.5 million are urban,
- 0.6 million are water, and
- 3.4 million are other lands (e.g., agricultural, etc.).

Substantially more acres of hardwood rangeland exist than reported in the 1981 photo-based mapping project. Most of the additional acreage is found on the north coast and north Sacramento valley. Aside from the omitted hardwood rangelands in the photo-based maps, both the photo-based and the image-based maps provide valuable information about California's hardwood rangelands.

The accuracy assessment of the 1981 and the 1990 maps gave map users valuable information:

- Both the photo-based and the image-based maps consistently slightly underestimate crown closure.
- Because certain hardwood species are difficult to distinguish from one another, cover type confusion can exist in areas where indistinguishable species occur together. It is difficult to distinguish the live oaks from one another on the ground, much less from remotely sensed data. In areas where both coast live oak and interior live oak occur, confusion in cover type labels will also occur. Similarly, valley oak on slopes is difficult to distinguish from blue oaks. Valley oak is the most confused type, and it is unlikely that mixed valley oak stands on slopes can be adequately classified from either aerial photography or satellite imagery. Finally, because of its sparse canopy, gray pine is extremely difficult to see on both aerial photography and satellite imagery. This lack of resolution in the remotely sensed data leads to confusion between blue oak woodland and blue oak–gray pine woodland.
- Errors of commission (labeling an area as hardwoods when it is not hardwoods) are rare in both maps and usually occur in mixed hardwood–conifer stands.
- Weakness in the classification system may create confusion because the system does not adequately handle classification of hardwood types that are dominated by interior live oak or Oregon white oak. In addition, many cover type indicators and associate tree species occur in several hardwood cover types. This is especially problematic in distinguishing the montane hardwood from the coast live oak cover type.

In addition, the assessment also showed that satellite image processing is a valuable tool for mapping and monitoring California's hardwood rangelands and

- Can produce maps that exceed the accuracy of maps from photo interpretation,
- Is flexible, allowing for changes in classification systems,
- Is cost-effective, allowing for a tripling of the project area extent with no increase in project cost,
- Provides increased detail of information and a richer database, and
- Facilitates monitoring of land use and land cover change over time.

CONCLUSIONS

The accuracy assessment analysis presented in this chapter shows both the complexity and subtle surprises that face any project. Several points are particularly important:

- Development and implementation of a robust classification system is critical to the success of any mapping or accuracy assessment project. As this chapter shows, the classification system plays a critical part in design, data collection, and analysis. Ambiguous systems without clear rules will result in a disastrous project.

- Analysis of accuracy assessment must go beyond the simple creation of an error matrix. The producer and user of the map need to understand why sites fall off of the diagonal.
- Is the confusion inevitable and *acceptable* as in that caused by variation in crown closure estimates, or is it inevitable and *unacceptable* as in that resulting from the misidentification of hardwood species?
- How much of the confusion is caused by reference error, overlapping classification system boundaries, inappropriate use of remote sensing technologies or map error? In this project,
  1. Reference error exists, especially in species identification. If photo interpretation of cover type is only 80% correct (Table 8-12), what is the impact on the map accuracy assessment, which relied heavily on photo interpretation for development of the reference data?
  2. Fuzzy class boundaries and variation in estimation were significant. Is this acceptable? In this case, cost savings of using remote sensing more than made up for the loss in map label precision. In other applications, the loss may be unacceptable, resulting in abandonment of remote sensing in favor of ground mapping.
- Trade-offs between sample design and data collection rigor and practical considerations are inevitable in most accuracy assessments. In particular, field data collection frequently precludes the use of random site selection. However, in each and every case it is important to remember that every method or procedure must be statistically valid as well as practically attainable. It is not acceptable to sacrifice either requirement; rather, the best balance must be found between the two.

References

Aickin, M. 1990. Maximum likelihood estimation of agreement in the constant predictive probability model, and its relation to Cohen's kappa. *Biometrics*. Vol. 46, pp. 293-302.

American Society of Photogrammetry. 1960. *Manual of Photographic Interpretation*. ASP, Washington, DC.

Anderson, J.R., E.E. Hardy, J.T. Roach, and R.E. Witner. 1976. A land use and land cover classification system for use with remote sensor data. USGS Professional Paper 964.

Aronoff, S. 1982. Classification accuracy: A user approach. *Photogrammetric Engineering and Remote Sensing*. Vol. 48, No. 8, pp. 1299-1307.

Aronoff, S. 1985. The minimum accuracy value as an index of classification accuracy. *Photogrammetric Engineering and Remote Sensing*. Vol. 51, No. 1, pp. 99-111.

Barrett, J.P., and M.E. Nutt. 1979. *Survey Sampling in the Environmental Sciences: A Computer Approach*. Compress, Inc., P. O. Box 102, Wentworth, NH.

Battesse, G., R. Harter, and W. Fuller. 1988. An error-component model for prediction of county crop areas using survey and satellite data. *Journal of the American Statistical Association*. Vol. 83, pp. 28-36.

Benson, A., and S. DeGloria. 1985. Interpretation of Landsat-4 Thematic Mapper and Multispectral Scanner data for forest surveys. *Photogrammetric Engineering and Remote Sensing*. Vol. 51, No. 9, pp. 1281-1289.

Berry, B.J.L. 1962. *Sampling, Coding, and Storing Flood Plain Data*. USDA Agriculture Handbook No. 237.

Berry, B.J.L., and A.M. Baker. 1968. Geographic sampling. *Spatial Analysis: A Reader in Statistical Geography*, B.J.L. Berry and D.F. Marble (Eds.). Prentice Hall, Englewood Cliffs, NJ, pp. 91-100.

Biging, G., and R. Congalton. 1989. Advances in forest inventory using advanced digital imagery. *Proceedings of Global Natural Resource Monitoring and Assessments: Preparing for the 21st Century*. Venice, Italy. September, 1989. Vol. 3, pp. 1241-1249.

Biging, G., R. Congalton, and E. Murphy. 1991. A comparison of photointerpretation and ground measurements of forest structure. *Proceedings of the 56th Annual Meeting of the American Society of Photogrammetry and Remote Sensing*. Baltimore, MD. Vol. 3, pp. 6-15.

Bishop, Y., S. Fienberg, and P. Holland. 1975. *Discrete Multivariate Analysis: Theory and Practice*. MIT Press, Cambridge, MA.

Brennan, R., and D. Prediger. 1981. Coefficient kappa: Some uses, misuses, and alternatives. *Educational and Psychological Measurement*. Vol. 41, pp. 687-699.

Buckland, S., and D. Elston. 1994. Use of groundtruth to correct land cover area estimates from remotely sensed data. *International Journal of Remote Sensing*. Vol. 15, No. 6, pp. 1273-1282.

Campbell, J.B. 1981. Spatial autocorrelation effects upon the accuracy of supervised classification of land cover. *Photogrammetric Engineering and Remote Sensing*. Vol. 47, No. 3, pp. 355-363.

Campbell, J.B. 1983. *Mapping the Land: Aerial Imagery for Land Use Information.* Association of American Geographers, Washington, DC.

Campbell, J. 1987. *Introduction to Remote Sensing.* Guilford Press, New York.

Card, D.H. 1980. Testing the accuracy of land use and land cover maps: A short review of the literature. NASA Ames Research Center.

Card, D.H. 1982. Using known map categorical marginal frequencies to improve estimates of thematic map accuracy. *Photogrammetric Engineering and Remote Sensing.* Vol. 48, No. 3, pp. 431-439.

Chrisman, N. 1982. Beyond accuracy assessment: Correction of misclassification. *Proceedings of the 5th International Symposium on Computer-Assisted Cartography.* Crystal City, VA, pp. 123-132.

Chuvieco, E., and R. Congalton. 1988. Using cluster analysis to improve the selection of training statistics in classifying remotely sensed data. *Photogrammetric Engineering and Remote Sensing.* Vol. 54, No. 9, pp. 1275-1281.

Cliff, A.D., and J.K. Ord. 1973. *Spatial Autocorrelation.* Pion, London.

Cochran, W.G. 1977. *Sampling Techniques.* John Wiley & Sons, New York.

Coggeshall, M.G., and R.M. Hoffer. 1973. Basic forest cover mapping using digitized remote sensor data and ADP techniques. Laboratory for Applications in Remote Sensing, Purdue University. LARS Information Note 030573.

Cohen, J. 1960. A coefficient of agreement for nominal scales. *Educational and Psychological Measurement.* Vol. 20, No. 1, pp. 37-40.

Cohen, J. 1968. Weighted kappa: Nominal scale agreement with provision for scaled disagreement or partial credit. *Psychological Bulletin.* Vol. 70, No. 4, pp. 213-220.

Colwell, R.N. 1955. The PI picture in 1955. *Photogrammetric Engineering.* Vol. 21, No. 5, pp. 720-724.

Congalton, R.G. 1981. The use of discrete multivariate analysis for the assessment of Landsat classification accuracy. M.S. Thesis, Virginia Polytechnic Institute and State University, Blacksburg, VA.

Congalton, R.G. 1984. A comparison of five sampling schemes used in assessing the accuracy of land cover/land use maps derived from remotely sensed data. Ph.D. Dissertation, Virginia Polytechnic Institute and State University, Blacksburg, VA.

Congalton, R.G. 1988a. Using spatial autocorrelation analysis to explore errors in maps generated from remotely sensed data. *Photogrammetric Engineering and Remote Sensing.* Vol. 54, No. 5, pp. 587-592.

Congalton, R.G. 1988b. A comparison of sampling schemes used in generating error matrices for assessing the accuracy of maps generated from remotely sensed data. *Photogrammetric Engineering and Remote Sensing.* Vol. 54, No. 5, pp. 593-600.

Congalton, R. 1991. A review of assessing the accuracy of classifications of remotely sensed data. *Remote Sensing of Environment.* Vol. 37, pp. 35-46.

Congalton, R., and G. Biging. 1992. A pilot study evaluating ground reference data collection efforts for use in forest inventory. *Photogrammetric Engineering and Remote Sensing.* Vol. 58, No. 12, pp. 1669-1671.

Congalton R., and K. Green. 1993. A practical look at the sources of confusion in error matrix generation. *Photogrammetric Engineering and Remote Sensing.* Vol. 59, No. 5, pp. 641-644.

Congalton, R.G., and R.A. Mead. 1983. A quantitative method to test for consistency and correctness in photo-interpretation. *Photogrammetric Engineering and Remote Sensing.* Vol. 49, No. 1, pp. 69-74.

Congalton, R., and R. Mead. 1986. A review of three discrete multivariate analysis techniques used in assessing the accuracy of remotely sensed data from error matrices. *IEEE Transactions of Geoscience and Remote Sensing*. Vol. GE-24, No 1, pp. 169-174.

Congalton, R.G., R.G. Oderwald, and R.A. Mead. 1983. Assessing Landsat classification accuracy using discrete multivariate statistical techniques. *Photogrammetric Engineering and Remote Sensing*. Vol. 49, No. 12, pp. 1671-1678.

Cowardin, L.M., V. Carter, F. Golet, and E. LaRoe. 1979. *A Classification of Wetlands and Deepwater Habitats of the United States*. Office of Biological Services. U.S. Fish and Wildlife Service, U.S. Department of Interior, Washington, DC.

Cox, K.R. 1969. The voting decision in a spatial context. *Progress in Geography*, L.C. Board, R.J. Charley, P. Haggett, and D.R. Stoddart (Eds.). Arnold, London, pp. 81-117.

Cruickshank, D.B. 1940. A contribution towards the rational study of regional influences: Group information under random conditions. *Papworth Research Bulletin*. Vol. 5, pp. 36-81.

Cruickshank, D.B. 1947. Regional influences in cancer. *British Journal of Cancer*. Vol. 1, pp. 109-128.

Czaplewski, R. 1992. Misclassification bias in aerial estimates. *Photogrammetric Engineering and Remote Sensing*. Vol. 58, No. 2, pp. 189-192.

Czaplewski, R., and G. Catts. 1990. Calibrating area estimates for classification error using confusion matrices. *Proceedings of the 56th Annual Meeting of the American Society for Photogrammetry and Remote Sensing*. Denver, CO. Vol. 4, pp. 431-440.

Eyre, F.H. 1980. *Forest Cover Types of the United States and Canada*. Society of American Foresters, Washington, DC.

Fienberg, S.E. 1970. An iterative procedure for estimation in contingency tables. *Annals of Mathematical Statistics*. Vol. 41, No. 3, pp. 907-917.

Fienberg, S.E. 1980. *The Analysis of Cross-Classified Categorical Data*. MIT Press: Cambridge, MA.

Fitzpatrick-Lins, K. 1978a. Accuracy of selected land use and land cover maps in the greater Atlanta region, Georgia. *Journal Research USGS*. Vol. 6, No. 2, pp. 169-173.

Fitzpatrick-Lins, K. 1978b. Accuracy and consistency comparisons of land use and land cover maps made from high-altitude photographs and Landsat MSS imagery. *Journal Research USGS*. Vol. 6, No. 1, pp. 23-40.

Fitzpatrick-Lins, K. 1981. Comparison of sampling procedures and data analysis for a land-use and land-cover map. *Photogrammetric Engineering and Remote Sensing*. Vol. 47, No. 3, pp. 343-351.

Fleiss, J. 1975. Measuring agreement between two judges on the presence or absence of a trait. *Biometrics*. Vol. 31, pp. 651-659.

Fleiss, J., J. Cohen, and B. Everitt. 1969. Large sample standard errors of kappa and weighted kappa. *Psychological Bulletin*. Vol. 72, No. 5, pp. 323-327.

Fleming, M.D., J.S. Berkebile, and R.M. Hoffer. 1975. Computer aided analysis of Landsat-1 MSS data: A comparison of three approaches, including a modified clustering approach. Laboratory for Applications in Remote Sensing, Purdue University. LARS Information Note 072475.

Foody, G. 1992. On the compensation for chance agreement in image classification accuracy assessment. *Photogrammetric Engineering and Remote Sensing*. Vol. 58, No. 10, pp. 1459-1460.

Freese, F. 1960. Testing accuracy. *Forest Science*. Vol. 6, No. 2, pp. 139-145.

Fung, T., and E. LeDrew. 1988. The determination of optical threshold levels for change detection using various accuracy indices. *Photogrammetric Engineering and Remote Sensing.* Vol. 54, No. 10, pp. 1449-1454.

Ghosh, S., J. Innes, and C. Hoffmann. 1995. Observer variation as a source of error in assessments of crown condition through time. *Forest Science.* Vol. 42, No. 2, pp. 235-254.

Ginevan, M.E. 1979. Testing land-use map accuracy: Another look. *Photogrammetric Engineering and Remote Sensing.* Vol. 45, No. 10, pp. 1371-1377.

Gong, P., and J. Chen. 1992. Boundary uncertainties in digitized maps: Some possible determination methods. IN: Proceedings of GIS/LIS '92. Annual Conference and Exposition. San Jose, CA. pp. 274-281.

Goodman, L. 1965. On simultaneous confidence intervals for multinomial proportions. *Technometrics.* Vol. 7, pp. 247-254.

Gopal, S., and C. Woodcock. 1994. Theory and methods for accuracy assessment of thematic maps using fuzzy sets. *Photogrammetric Engineering and Remote Sensing.* Vol. 60, No. 2, pp. 181-188.

Grassia, A., and R. Sundberg. 1982. Statistical precision in the calibration and use of sorting machines and other classifiers. *Technometrics.* Vol. 24, pp. 117-121.

Gregg, T.W.D., E.W. Barthmaier, R.E. Aulds, and R.B. Scott. 1979. Landsat operational inventory study. Division of Technical Services, State of Washington Department of Natural Resources, Olympia, WA.

Griffin, J., and W. Critchfield. 1972. The distribution of forest trees in California. USDA Forest Service Research Paper PSW-82. Pacific Southwest Forest and Range Experiment Station, Berkeley, CA.

Hay, A.M. 1979. Sampling designs to test land-use map accuracy. *Photogrammetric Engineering and Remote Sensing.* Vol. 45, No. 4, pp. 529-533.

Hay, A.M. 1988. The derivation of global estimates from a confusion matrix. *International Journal of Remote Sensing.* Vol. 9, pp. 1395-1398.

Hill, T.B. 1993. Taking the " " out of "ground truth": Objective accuracy assessment. IN: Proceedings of the 12th Pecora Conference. Sioux Falls, SD. pp. 389-396.

Hixson, M., D. Scholz, N. Fuhs, and T. Akiyama. 1980. Evaluation of several schemes for classification of remotely sensed data. *Photogrammetric Engineering and Remote Sensing.* Vol. 46, No. 12, pp. 1547-1553.

Hoffer, R.M. 1975a. Computer aided analysis of skylab multispectral scanner data in mountainous terrain for land use, forestry, water resources, and geologic applications. Laboratory for Applications in Remote Sensing, Purdue University. LARS Information Note 121275.

Hoffer, R.M. 1975b. Natural resource mapping in mountainous terrain by computer analysis of ERTS-1 satellite data. Laboratory for Applications in Remote Sensing, Purdue University. LARS Research Bulletin 919. LARS Information Note 061575.

Hoffer, R.M., and M.D. Fleming. 1978. Mapping vegetative cover by computer aided analysis of satellite data. Laboratory for Applications in Remote Sensing, Purdue University. LARS Technical Report 011178.

Hord, R.M., and W. Brooner. 1976. Land-use map accuracy criteria. *Photogrammetric Engineering and Remote Sensing.* Vol. 42, No. 5, pp. 671-677.

Hudson, W., and C. Ramm. 1987. Correct formulation of the kappa coefficient of agreement. *Photogrammetric Engineering and Remote Sensing.* Vol. 53, No.4, pp. 421-422.

Kalensky, Z., and L. Scherk. 1975. Accuracy of forest mapping from Landsat computer compatible tapes. *10th International Symposium on Remote Sensing of Environment,* Ann Arbor, MI, pp. 1159-1167.

Katz, A.H. 1952. Photogrammetry needs statistics. *Photogrammetric Engineering.* Vol. 18, No. 3. pp. 536-542.

Kish, L. 1965. *Survey Sampling.* John Wiley & Sons, New York.

Landgrebe, D.A. 1973. Machine processing for remotely acquired data. Laboratory for Applications in Remote Sensing, Purdue University. LARS Information Note 031573.

Landis, J., and G. Koch. 1977. The measurement of observer agreement for categorical data. *Biometrics.* Vol. 33, pp. 159-174.

Law, A.M., and W.D. Kelton. 1982. *Simulation Modeling and Analysis.* McGraw-Hill, New York.

Learmonth, G.P. 1976. Empirical tests of multipliers for the prime modulus random number generator (x(i+1) = AX(i) mod 2(31)-1). *Proceedings of the 9th Interface Symposium on Computer Science and Statistics.* Prindle, Weber, and Schmidt, Boston, pp. 178-183.

Lillesand, T.M., and R.W. Kiefer. 1979. *Remote Sensing and Image Interpretation.* John Wiley & Sons, New York.

Ling, H.S., G.H. Rosenfield, and K. Fitzpatrick-Lins. 1979. Sample selection for thematic map accuracy testing. USGS circular.

Loveland, T.R., and G.E. Johnson. 1983. The role of remotely sensed and other spatial data for predictive modeling—the Umatilla, Oregon example. *Photogrammetric Engineering and Remote Sensing.* Vol. 49, No. 8, pp. 1183-1197.

Lowell, K. 1992. On the incorporation of uncertainty into spatial data systems. IN: Proceedings of GIS/LIS '92. Annual Conference and Exposition. San Jose, CA. pp. 484-493.

Lunetta, R., R. Congalton, L. Fenstermaker, J. Jensen, K. McGwire, and L. Tinney. 1991. Remote sensing and geographic information system data integration: error sources and research issues. *Photogrammetric Engineering and Remote Sensing.* Vol. 57, No. 6, pp. 677-687.

Lyon, J.G. 1978. An analysis of vegetation communities in the lower Columbia river basin. *Pecora IV Symposium.* Sioux Falls, SD.

Lyon, J.G. 1979. Remote sensing analyses of coastal wetlands characteristics: The St. Clair Flats, Michigan. *13th International Symposium on Remote Sensing of Environment.* Ann Arbor, MI. pp. 1117-1129.

Malila, W. 1985. Comparison of the information contents of Landsat TM and MSS data. *Photogrammetric Engineering and Remote Sensing.* Vol. 51, No. 9, pp. 1449-1457.

Martin, L.R.G. 1989. Accuracy assessment of Landsat-based visual change detection methods applied to the rural-urban fringe. *Photogrammetric Engineering and Remote Sensing.* Vol. 55, pp. 209-215.

Mayer, K., and W. Laudenslayer (Eds.). 1988. A guide to wildlife habitats of California. California Department of Forestry and Fire Protection, Sacramento, CA.

McGuire, K. 1992. Analyst variability in labeling unsupervised classifications. *Photogrammetric Engineering and Remote Sensing.* Vol. 58, No. 12. pp. 1705-1709.

Mead, R.A., and M.P. Meyer. 1977. Landsat digital data application to forest vegetation and land use classification in Minnesota. *Machine Processing of Remotely Sensed Data, Proceedings.* Purdue University. pp. 270-280.

Meyer, M., and L. Werth. 1990. Satellite data: Management panacea of potential problem? *Journal of Forestry.* Vol. 88, No. 9, pp. 10-13.

Meyer, M., J. Brass, B. Gerbig., and F. Batson. 1975. ERTS data applications to surface resource surveys of potential coal production lands in southeast Montana. IARSL Research Report 75-1. Final Report. University of Minnesota.

Moran, P.A.P. 1948. The interpretation of statistical maps. *Journal Royal Stat. Soc.,* Series B. Vol. 10, pp. 243-251.

Ott, L. 1977. *An Introduction to Statistical Methods and Data Analysis.* Duxbury Press, North Scituate, MA.

Pacific Meridian Resources. 1994. California hardwood rangeland monitoring: Final Report. California Department of Forestry and Fire Prevention, Sacramento, CA.

Pillsbury, N., M. DeLasaux, R. Pryor, and W. Bremer. 1991. Mapping and GIS database development for California's hardwood resources. California Department of Forestry and Fire Protection, Forest and Rangeland Resources Assessment Program (FRRAP), Sacramento, CA.

Prisley, S., and J. Smith. 1987. Using classification error matrices to improve the accuracy of weighted land-cover models. *Photogrammetric Engineering and Remote Sensing.* Vol. 53, No. 9, pp. 1259-1263.

Pritsker, A.A.B., and C.D. Pegden. 1979. *Introduction to Simulation and SLAM.* John Wiley & Sons, New York.

Quenouille, M.H. 1949. Problems in plane sampling. *Annals of Mathematical Statistics.* Vol. 20, pp. 355-375.

Reed, D.D. 1982. The spatial autocorrelation of individual tree characteristics in Loblolly Pine stands. Ph.D. Dissertation. Virginia Polytechnic Institute and State University.

Reitman, J. 1971. *Computer Simulation Applications: Discrete-Event Simulation for the Synthesis and Analysis of Complex Systems.* John Wiley & Sons, New York.

Rhode, W. 1977. Digital image analysis techniques required for natural resource inventories. *AFIPS Conference Proceedings.* Vol. 47, pp. 93-106.

Rhode, W.G. 1978. Digital image analysis techniques for natural resource inventories. *National Computer Conference Proceedings,* pp. 43-106.

Rhode, W.G., and W.A. Miller. 1980. Arizona vegetation resource inventory (AVRI) project final report. USGS EROS Data Center internal report to BLM. Sioux Falls, SD.

Rosenfield, G.H. 1980. Analysis of variance of thematic mapping experiment data. *Proceedings of the American Society of Photogrammetry Annual Conference.* St. Louis, MO.

Rosenfield, G. 1981. Analysis of variance of thematic mapping experiment data. *Photogrammetric Engineering and Remote Sensing.* Vol. 47, No. 12, pp. 1685-1692.

Rosenfield, G.H. 1982. Sample design for estimating change in land use and land cover. *Photogrammetric Engineering and Remote Sensing.* Vol. 48, No. 5, pp. 793-801.

Rosenfield, G., and K. Fitzpatrick-Lins. 1986. A coefficient of agreement as a measure of thematic classification accuracy. *Photogrammetric Engineering and Remote Sensing.* Vol. 52, No. 2, pp. 223-227.

Rosenfield, G.H., and M.L. Melley. 1980. Applications of statistics to thematic mapping. *Photogrammetric Engineering and Remote Sensing.* Vol. 46, No. 10, pp. 1287-1294.

Rosenfield, G.H., K. Fitzpatrick-Lins, and H. Ling 1982. Sampling for thematic map accuracy testing. *Photogrammetric Engineering and Remote Sensing.* Vol. 48, No. 1, pp. 131-137.

Rubinstein, R.Y. 1981. *Simulation and the Monte Carlo Method.* John Wiley & Sons, New York.

Rudd, R.D. 1971. Macro land use mapping with simulated space photographs. *Photogrammetric Engineering.* Vol. 37, pp. 365-372.

Sammi, J.C. 1950. The application of statistics to photogrammetry. *Photogrammetic Engineering.* Vol. 16, No. 5. pp. 681-685.

Shasby, M.B., R.E. Burgan, and G.R. Johnson. 1981. Broad area forest fuels and topography mapping using digital Landsat and terrain data. *Proceedings of the 7th International Symposium on Machine Processing of Remotely Sensed Data.* Purdue University, West Lafayette, IN, pp. 529-538.

Singh, A. 1986. Change detection in the tropical rain forest environment of Northeastern India using Landsat. In *Remote Sensing and Tropical Land Management.* Edited by Eden, M.J., and Parry, J.T., London: John Wiley & Sons, pp. 237-254.

Skidmore, A., and B. Turner. 1989. Assessing the accuracy of resource inventory maps. *Proceedings of Global Natural Resource Monitoring and Assessments: Preparing for the 21st Century.* Venice, Italy, September, 1989. Vol. 2, pp. 524-535.

Skidmore, A., and B. Turner. 1992. Map accuracy assessment using line transect sampling. *Photogrammetric Engineering and Remote Sensing.* Vol. 58, No. 10, pp. 1453-1457.

Smiatek, G. 1995. Sampling Thematic Mapper imagery for land use data. *Remote Sensing of Environment.* Vol. 52, pp. 116-121.

Snedecor, G.W., and W.G. Cochran. 1976. *Statistical Methods.* 6th ed. Iowa State Press, Ames, IA.

Sobol, I.M. 1974. *The Monte Carlo Method.* University of Chicago Press, Chicago.

Spurr, S. 1948. *Aerial Photographs in Forestry.* Ronald Press, New York.

Stehman, S. 1992. Comparison of systematic and random sampling for estimating the accuracy of maps generated from remotely sensed data. *Photogrammetric Engineering and Remote Sensing.* Vol. 58, No. 9, pp. 1343-1350.

Stenlund, H., and A. Westlund. 1975. A Monte Carlo study of simple random sampling from a finite population. *Scandinavian Journal of Statistics.* Vol. 2, pp. 106-108.

Story, M., and Congalton, R. 1986. Accuracy assessment: A user's perspective. *Photogrammetric Engineering and Remote Sensing.* Vol. 52, No. 3, pp. 397-399.

Tenenbein, A. 1972. A double sampling scheme for estimating from misclassified multinomial data with applications to sampling inspection. *Technometrics.* Vol. 14, pp. 187-202.

Todd, W., D. Gehring, and J. Haman. 1980. Landsat wildland mapping accuracy. *Photogrammetric Engineering and Remote Sensing.* Vol. 43, No. 9, pp. 1135-1137.

Tortora, R. 1978. A note on sample size estimation for multinomial populations. *The American Statistician.* Vol. 32, No. 3, pp. 100-102.

van Genderen, J.L., and B.F. Lock. 1977. Testing land use map accuracy. *Photogrammetric Engineering and Remote Sensing.* Vol. 43, No. 9, pp. 1135-1137.

van Genderen, J.L., B.F. Lock, and P.A. Vass. 1978. Remote sensing: statistical testing of thematic map accuracy. *Proceedings of the 12th International Symposium on Remote Sensing of Environment,* ERIM, Ann Arbor, MI, pp. 3-14.

Woodcock, C. 1996. On roles and goals for map accuracy assessment: A remote sensing perspective. *2nd International Symposium on Spatial Accuracy Assessment in Natural Resources and Environmental Sciences,* USDA Forest Service Rocky Mountain Forest and Range Experiment Station, Gen. Tech. Rep. RM-GTR-277. Fort Collins, CO, pp. 535-540.

Woodcock, C., and S. Gopal. 1992. Accuracy assessment of the Stanislaus Forest vegetation map using fuzzy sets. *Remote Sensing and Natural Resource Management. Proceedings of the 4th Forest Service Remote Sensing Conference,* Orlando, FL, pp. 378-394.

Yates, F. 1981. *Sampling Methods for Censuses and Surveys.* Charles Griffin, London.

Young, H.E. 1955. The need for quantitative evaluation of the photo interpretation system. *Photogrammetric Engineering.* Vol. 21, No. 5. pp. 712-714.

Young, H.E., and E.G. Stoeckler. 1956. Quantitative evaluation of photo interpretation mapping. *Photogrammetric Engineering.* Vol. 22, No. 1. pp. 137-143.

Zadeh, L.A. 1965. Fuzzy sets. *Information and Control.* Vol. 8, pp. 338-353.

Zonneveld, I.S. 1974. Aerial photography, remote sensing and ecology. *ITC Journal.* Part 4, pp. 553-560.

Index

A

Accuracy, types of, 2–3
Accuracy assessment, See also Error matrices
  case study, 85–122, See California hardwood rangeland monitoring project
  cutoff between acceptable and unacceptable accuracy, 45
  fundamental steps for, 4
  history of, 8–10
  multilayer assessments, 83–84
  rationale for, 3–4
  reference data collection, 27–41, See Data collection
  sampling, 11, See Sample design
  statistical analysis, See Analysis
Aerial photography, See also Photo interpretation
  California hardwood rangeland monitoring project, 85–122
  history of photo interpretation, 7–8
Analysis techniques, 43–64
  area estimation/correction (calibration), 63–64
  assumptions regarding sampling scheme, 25
  California hardwood rangeland monitoring project, 99–121
  compensation for chance agreement, 58
  conditional Kappa, 56–57
  confidence limits, 59–63
  discrete versus continuous data, 14
  Kappa analysis, 49–53, 105
  non-site specific assessments, 43–44
  normalized accuracy using iterative proportional fitting (Margfit), 53–56
  weighted Kappa, 57–58
Autocorrelation, 14–16, 89, 93

B

Binomial distribution, for sample size computation, 18, 19

C

California Department of Forestry and Fire Protection, 77, 86
California hardwood rangeland monitoring project, 85–122
  classification scheme, 89–90
  cover type analysis, 114–120
  crown closure analysis, 105–114
  data collection, 95–99
  error matrix development, 99–104
  errors of omission and commission, 110, 122
  extent, 120–121
  findings and implications, 121–123
  quality control, 98
  sample design, 86–95
  sample units, 89–90
  sampling scheme, 93–95
  statistical analysis, 105
California Polytechnic Institute, 86
Chance agreement, 58
Change detection, 80–83
Classical estimator, 63
Classification accuracy, 2–3
Classification scheme, 11–14, 87, 122

appropriateness of remote sensing
technology, 69
California hardwood rangeland
monitoring project, 87–89
complex schemes, 75
errors, 66
fuzzy rules, 75, 76–79
land cover mapping classification key,
71–73
non-site specific assessments, 43–44
sensitivity to observer variability, 67–68
Classified data, 9
Cluster sampling, 23
Commission errors, 9–10, 45, 110, 122
Conditional Kappa, 56–57
Confidence limits, 59–63
Confusion matrix, 116, See also Fuzzy set
theory
Consistency, 38–40
Continuous data, 14
fuzzy set theory, 76–79
variation in human interpretation and, 75
Cover classification, See Land cover
classification; Classification scheme
Crown closure, 14, 29, 30
California hardwood rangeland
monitoring project, 100, 105–114
difference matrices, 101–102, 106
error matrix for field measurement versus
visual call, 34–35
fuzzy rules, 78–79
interpreter variability, 68
labels, California hardwood rangeland
monitoring project, 100
reference data collection, 29, 30, 31–33

D

Data collection, 27–41
California hardwood rangeland
monitoring project, 95–99
collection method, 30–33, 98
consistency, 38–40
data independence, 37–38
errors, 41
objectivity and consistency, 36–41
observations versus measurements,
30–31, 38
observer variability, 31, See also
Observer variability

photos versus ground, 29
quality control, 40–41
reference data collection form, 38–40
source data, 28–30, 95
timeliness, 35–36, 98
Data consistency, 38–40, 98
Data entry errors, 41, 66, 99
Data independence, 37–48, 98
Delta method, 50
Difference image, 15
Difference matrices, 100–107
normalized, 106–107
Digital accuracy assessment, history of,
7, 8–10
Discrete data, 14

E

Error matrices, 9–10, 45–48, 51–53
accounting for fuzzy class boundaries,
76–79, See also Fuzzy set theory
California hardwood rangeland
monitoring project, 99–121
change detection, 81–82
comparing using Kappa analysis, 25, See
Kappa analysis
complex classification schemes and, 75
confidence intervals, 59–63
extent, 120–121
field measurement versus visual ground
cover calls, 34–35
initial (difference) matrices, 100–104
map labels, 9, 100
reference label differences, 65–70
mapping error and, 69–70
mathematical representation of, 47–48
no-change/change, 82–83
normalization using Margfit iterative
proportional fitting procedure, 53–56
overall accuracy, 10, 45–46, 48, 61
producer's accuracy, 10, 46, 48, 60, 62
reference data errors and, 65–66
remote sensing technology
appropriateness, 69
sample size, 18–22
sensitivity to observer variability, 67–68,
See also Observer variability
statistical analysis, See Analysis
user's accuracy, 10, 46, 48, 60, 62
updating category areal estimates, 63
zeros in, 55

Errors of exclusion (omission errors), 9–10, 45, 110, 122
Errors of inclusion (commission errors), 9–10, 45, 110, 122

F

Field-based photo interpretation accuracy, 105, 108–110
Field inventory, 31–33
Fire, 35, 114
Forest, defining, 12
Forest density, See Crown closure
Fuel vegetation, 13
Fuzzy set theory (fuzzy logic), 63, 68, 75, 76–79, 100, 114, 116

G

Global positioning system (GPS), 16
Graminoid cover, 70
Ground-based reference data collection, 28, 29, 31–33

H

Hardwood monitoring project, See California hardwood rangeland monitoring project
Harvests, 35, 114
Hierarchical classification scheme, 13

I

Inter-interpreter variability, See Observer variability
Inverse estimator, 63
Iterative proportional fitting (Margfit) procedure, 53–56, 105

K

Kappa analysis, 25, 49–53
California hardwood rangeland monitoring project, 105, 108
compensation for chance agreement, 58
conditional Kappa, 56–57
weighted Kappa, 57–58
KHAT statistic, 49–55

L

Land cover changes, 66, 114
detection of, 80–83
Land cover classification, 12, See also Classification scheme
California hardwood rangeland monitoring project, 114–120
classification key, 71–73
density, See Crown closure
difference matrix, 103–104, 107
discrete vs. continuous data, 14, 67
reference data, See Data collection; Reference data
tree size class, 29–34
water, 69–70
Landsat 1, 8
Landsat MSS data, 14–15
Landsat Thematic Mapper data, 18, 29, 35–36, 51, 53, 69
Live oak type distinctions, 13, 122
Location accuracy, 2

M

Map labels, 9, 12, 100
Map marginal proportion, 59
Mapping error, 69–70
Margfit, 53–56, 105
Marginal proportions, 59–60
Multilayer assessments, 83–84
Multinomial distribution for sample size determination, 18–22

N

National Wetlands Inventory (NWI), 66
No-change/change error matrix, 82–83
Non-site specific assessments, 43–44
Normalized binomial distribution, 18, 19
Normalized difference matrices, 106–107
Normalized matrices (Margfit iterative proportional fitting), 53–56, 105

O

Objectivity and consistency, 36–41
Observer variability, 75
classification scheme complexity and, 75

classification scheme sensitivity to,
     67–68
measuring, 79–80
vegetation cover reference data
     collection, 31
Omission errors, 9–10, 45, 110, 122
Overall accuracy, 10, 45–46
     confidence interval computation, 61
     mathematical formula, 48

P

Pacific Meridian Resources, 77, 86
Personnel training, 38
Photo interpretation
     accuracy, field versus office, 96, 105,
          108–110
     ground-based reference data versus, 29
     history of quantitative evaluation, 7–8
     variation in human interpretation, 67–68,
          75, 79–80, See also Observer
          variability
Pixel, 16
     change detection, 80
     sample size selection, 18
     sample unit, 16
Plato, 67
Polygon sample units, 16–17, 89, 94, 114
Producer's accuracy, 10, 46
     confidence interval calculation, 60, 62
     mathematical formula, 48

Q

Quality control, 40–41, 66
Quinalt Indian Reservation, 77

R

Random sampling, 23–25, 94
Reference data
     errors in, 65–66
     sampling, 11, See Sample design
     collection, 27–41, See Data collection
     collection forms, 38–40
Reference site, 9
     label, 100
     location, 40
Remote sensing

appropriateness for specific applications,
     69
history of aerial photography, 7–8
rationale for using data from, 2–3
sources of error in data, 3, 81
Roads, 25

S

Sample design
     appropriate sample unit, 16–17, 89–90
     California hardwood rangeland
          monitoring project, 86–85
     classification schemes, 11–14, 87, 122
     distribution of map categories, 14, 87
     number of samples (sample size), 17–22,
          91–92
     sample selection (sampling scheme),
          22–25, 93–95
     spatial autocorrelation and, 14–16
Sample location, 40
Sample size, 17–22
Sample units, 16–17
     California hardwood rangeland
          monitoring project, 89–90
     delineation, 27, 40
Sampling, 11, See Sample design; Sampling
     scheme
Sampling scheme, 22–25
     California hardwood rangeland
          monitoring project, 93–95
Satellite imagery
     California hardwood rangeland
          monitoring project, 85–122
     Landsat MSS data, 14–15
     Landsat TM data, 18, 29, 35–36, 51, 53,
          69
     SPOT, 35
Simple random sampling, 23–25
Site specific assessment, 9
     basic analysis techniques, 45–63,
          See Analysis techniques
Spatial autocorrelation, 14–16, 89, 93
SPOT, 35
Spotted owl habitat, 30–31
Spruce, 69
Statistical analysis, 43–64, See Analysis
     techniques
Stratified random sampling, 23, 24–25
Stratified systematic unaligned sampling, 23
Systematic sampling, 23

T

Thematic accuracy, 2–3
Tongass National Forest, 77
Training, 38
Transition zones, 76
    spatial autocorrelation, 15
Tree size class, 29–34

U

Urban expansion, 114
U.S. Environmental Protection Agency
    (EPA), 12
User's accuracy, 10, 46
    confidence interval calculation, 60, 62
    mathematical formula, 48
U.S. Forest Service, 12, 77

V

Variance labels, 100
Variation in human interpretation, 67–68,
    75, 79–80, See also Observer variability

W

Water mapping, 69–70
Weighted Kappa, 57–58
Wetlands, 66
Wrangell-St. Elias National Park and
    Preserve, 68, 69, 70, 71–73
Wright, Wilbur, 7